T0329324

Electronic Waste

Acknowledgment

I would like to dedicate this book to my wife Dr. Mary Sexton for her unwavering support.

Electronic Waste
Toxicology and Public Health Issues

Bruce A. Fowler

ELSEVIER

ACADEMIC PRESS
An imprint of Elsevier

Academic Press is an imprint of Elsevier
125 London Wall, London EC2Y 5AS, United Kingdom
525 B Street, Suite 1800, San Diego, CA 92101-4495, United States
50 Hampshire Street, 5th Floor, Cambridge, MA 02139, United States
The Boulevard, Langford Lane, Kidlington, Oxford OX5 1GB, United Kingdom

Notices
Knowledge and best practice in this field are constantly changing. As new research and experience broaden our understanding,
changes in research methods, professional practices, or medical treatment may become necessary.

Practitioners and researchers must always rely on their own experience and knowledge in evaluating and using any
information, methods, compounds, or experiments described herein. In using such information or methods they should be
mindful of their own safety and the safety of others, including parties for whom they have a professional responsibility.

To the fullest extent of the law, neither the Publisher nor the authors, contributors, or editors, assume any liability for any
injury and/or damage to persons or property as a matter of products liability, negligence or otherwise, or from any use or
operation of any methods, products, instructions, or ideas contained in the material herein.

Library of Congress Cataloging-in-Publication Data
A catalog record for this book is available from the Library of Congress

British Library Cataloguing-in-Publication Data
A catalogue record for this book is available from the British Library

ISBN: 978-0-12-803083-7

For information on all Academic Press publications visit our website at
https://www.elsevier.com/books-and-journals

Working together
to grow libraries in
developing countries

www.elsevier.com • www.bookaid.org

Acquisition Editor: Erin Hill-Parks
Editorial Project Manager: Tracy Tufaga
Production Project Manager: Edward Taylor
Designer: Ines Cruz

Typeset by TNQ Books and Journals

Contents

Biography

Bruce A. Fowler PhD, Fellow ATS

Education

B.S. degree in Fisheries (Marine Biology)—University of Washington in 1968
PhD in Pathology—University of Oregon Medical School in 1972

Dr. Fowler began his scientific career at the National Institute of Environmental Health Sciences prior to becoming Director of the University of Maryland System-wide Program in Toxicology and Professor at the University of Maryland School of Medicine. He then served as Associate Director for Science in the Division of Toxicology and Environmental Medicine at Agency for Toxic Substances and Disease Registry (ATSDR). He is currently a private consultant and Co-owner of Toxicology Risk Assessment Consulting Services (TRACS), LLC. In addition, Dr. Fowler serves as an Adjunct Professor, Emory University Rollins School of Public Health, and Presidents Professor of Biomedical Sciences, Center for Alaska Native Health Research (CANHR) at the University of Alaska Fairbanks. Dr. Fowler is an internationally recognized expert on the toxicology of metals and has served on a number of state, national, and international committees in his areas of expertise. These include the Maryland Governor's Council on Toxic Substances (Chair), various National Academy of Sciences/National Research Council Committees, the USEPA Science Advisory Board, and Fulbright Scholarship review committee for Scandinavia (Chair, 2000–01). In 2016, he became an Inaugural Member of the Fulbright 1946 Society. He has also served as a temporary advisor to the World Health Organization (WHO) and the International Agency for Research Against Cancer (IARC) for a number of toxicology and risk assessment issues and has been recently appointed as a member of the Joint FAO/WHO Expert Committee on Food Additives (JECFA) for the period 2016–20. Dr. Fowler has been honored as a Fellow of the Japanese Society for the Promotion of Science (JSPS), a Fulbright Scholar, and Swedish Medical Research Council Visiting Professor at the Karolinska Institute, Stockholm, Sweden, and elected as a Fellow of the Academy of Toxicological Sciences. His more recent awards include a CDC/ATSDR, Honor Award for Excellence in Leadership Award 2010, and The USP Toxicology Committee 2010–15. The USP Elemental Impurities Panel and the 2014 US Pharmacopeia Award for an Innovative Response to Public Health Challenges (Group Award). He was appointed to the USP Nanotechnology Subcommittee in 2015. Dr. Fowler was previously elected to the Council of the Society of Toxicology (2005–07), the Board of Directors of the Academy of Toxicological Sciences (2006–09), and more recently, to the Council of the Society for Risk Analysis (2014–17). He is the Federal Legislative and National Active and Retired Federal Employees Association (NARFE)-PAC Chair for the Rockville Maryland Chapter of NARFE. Dr. Fowler is the current President of the Rotary Club of North Bethesda, Maryland (2016–17), and was selected as Rotarian of the Year in 2015 for his work in developing a taxi-based program to help persons with disabilities gain independence via reliable transportation to work. Dr. Fowler is the author of over 260 research papers and book chapters dealing with molecular mechanisms of metal toxicity, molecular biomarkers for early detection of metal-induced cell injury, and application of computational toxicology for risk assessment. He has been the editor, coeditor, or author of eight books or monographs on metal toxicology and mechanisms of chemical-induced cell injury, molecular biomarkers, and risk assessment and computational toxicology. Dr. Fowler currently focuses on the global problem of electronic waste in developing countries. He serves on the editorial boards of a number of scientific journals in toxicology and is an Associate Editor of the journal *Toxicology and Applied Pharmacology* and a past Associate Editor of *Environmental Health Perspectives* (2007–16).

Current Research Interests:

Toxicology of metals
Application of molecular biomarkers for risk assessment
Mechanisms of cell injury/cell death
Computational toxicology and risk assessment
Global impacts of electronic waste in developing countries

RECENT PUBLICATIONS

- McPhail B, Tie Y,Hong, H, Pearce BA, Schnackenberg L, Valerio Jr. LG, Fuscoe JC, TongW, Buzatu DA, Wilkes JG, Fowler BA, Demchuk E, Beger RD. Modeling interaction profiles: I. Spectral data-activity relationship and structure- activity relationship models for inhibitors and non-inhibitors of cytochrome P450 CYP3A4 and CYP2D6 isozymes. Molecules: QSAR and Its Applications. 17:3383–3406, 2012.

- Tie Y, McPhail B, Hong H, Pearce BA, Schnackenberg L, Weigong G, Buzatu DA, Wilkes JG, Fuscoe JC, Tong W, Fowler BA, Beger RD, Demchuk E. Modeling chemical interaction profiles: II. Molecular docking, spectralactivity relationship, and structure activity relationship models forpotent and weak inhibitors of cytochrome P450 CYP3A4 isozyme. Molecules: QSAR and Its Applications. 17:3407–3460, 2012.
- Fowler BA. Biomarkers in toxicology and risk assessment. In: Volume 3. Environmental toxicology. Luch, A. (ed) Birkhauser-Springer. Experientia Volume 101: 459–470, 2012.
- Fowler BA (ed). Computational toxicology: Applications for risk assessment. Elsevier Publishers, Amsterdam (2013) pp 258. (Book).
- Nordberg GF, Fowler BA Nordberg M (eds.). Handbook on the toxicology of metals (4th Edition) Elsevier Publishers, Amsterdam (2015) pp 1385. (Book).
- Go YM, Sutliff RL, Chandler JD, Khalidur R, Kang BY, Anania FA, Orr M, Hao L, Fowler BA, Jones DP. Low-dose cadmium causes metabolic and genetic dysregulation associated with fatty liver disease in mice. Toxicol Sci. 247:524–534, 2015.
- Fowler BA. Molecular biological markers for toxicology and risk assessment. Elsevier Publishers, Amsterdam (2016) 153 pp. (Book).
- Ruiz P, Perlina A, Mumtaz M, Fowler BA. A systems biology approach reveals converging molecular mechanisms that link different POPs to common metabolic diseases. Environ. Health Perspect. 124(7):1034–1041, 2016.

Foreword

This is the third book in a series that considers the utilization of the tools of modern molecular toxicology and computational toxicology to address pressing cutting edge public health issues in an integrated manner. The rapid global expansion of the electronic industry over the past 30 years has resulted in the production of many useful devices such as computers, printers, cell phones, tablets, and televisions, which have brought many benefits to mankind. Many of these electronic products contain inorganic and organic toxic chemicals to which humans may be exposed during both the fabrication and recycling/disposal phases of these devices if suitable precautions are not taken. This book will focus on the growing global electronic waste (e-waste) aspects of this problem area as a public health issue on various interrelated levels. It will consider various components of e-waste starting from macroscale with the millions of tons/year of outdated electronic products becoming available, through current recycling approaches in both developed and developing countries including child/gender labor force issues, existing epidemiological studies, and known mechanisms of toxicity for some of the major inorganic and organic on both an individual and chemical mixtures basis. The book will examine currently available risk assessments for e-waste. It will suggest some new approaches for addressing public health problems associated with this industry including the use of molecular biomarkers for early detection of toxicity and subpopulations at special risk. It will conclude with the application of computational modeling approaches for both evaluating relatively unstudied chemicals/chemical mixtures and processing molecular biomarker data for incorporation into new systems biology–based risk assessments. The overall goal of this book, from a scientific perspective, is hence to promote the use of 21st-century tools to address a unique and rapidly growing global public health issue related to the disposal of high technology electronic devices. In so doing, this book will demonstrate how the studies reviewed in two prior books on computational toxicology (Fowler, 2013) and molecular biomarkers (Fowler, 2016), respectively, may be effectively used to address a major global public health problem linked to high technology electronic products. It is hoped that, overall, this three-volume series will be a useful and integrated resource for academics, students, risk assessors, and regulators who are engaged in handling e-waste issues now and in the future.

Bruce A. Fowler PhD, ATS
Emory University School of Public Health and
University of Alaska—Fairbanks

Preface

The production of small electronic devices including televisions, personal computers, cell phones, and printers have brought many societal benefits over the past 50 years, but massive accumulation of these devices has generated major public health concerns due to the presence toxic chemicals in them. Too often inadequate disposal/recycling practices have exacerbated the concerns. This problem is of particular concern in developing countries with emerging economies, weak public health infrastructures, and rapidly growing populations. This book on electronic e-waste will cover the magnitude of the problem on a global basis, current recycling practices, the toxicology of some of the more common chemicals found in e-waste, and the human populations most commonly exposed to these agents. Particular emphasis will be placed on the transference of e-waste to developing countries where populations of concern include children working in recycling activities and impoverished groups with poor nutritional status and limited access to medical resources. The issue of exposure to chemical mixtures will be reviewed since this issue is a central feature of e-waste due to the presence of a number of organic and inorganic chemicals in modern electronic devices.

CHAPTER 1

Magnitude of the Global E-Waste Problem

1. SCOPE OF THE PROBLEM

1.1 Electronic Waste Types

There are a number of new electronic devices introduced into commerce on a daily basis to replace older existing produces (Greenpeace, 2009; ILO, 2012; UNEP, 2009; Wikipedia, 2016). The rate at which these new machines are being produced is accelerating since they are being produced in virtually every country of the world (ILO, 2012). The net result of this process is a surplus of older electronic machines that are entering the waste stream on a global basis in enormous quantities (UNEP, 2009). There also appears to be a large backlog of electronic devices currently stored in homes, which have not entered into the e-waste stream (Saphores, Nixon, Ogunseitan, & Shapiro, 2009). The types of electronics that become e-waste include but are not limited to the following: computers, printers, cellular telephones, televisions, washing machines, refrigerators, and more recently, "smart" electronic devices such as tablets and smartphones. Molecular robots with artificial intelligence are also in development (Hagiya, Konagaya, Kobayashi, Saito, & Murata, 2014). All of these useful devices have a finite lifetime, which is increasingly shortened by the advent of newer and faster and more intelligent machines (Singh, Li, & Zeng, 2016). Electronic devices contain a number of toxic chemicals, some of which are highly valuable such as gold, indium, gallium, and copper and worthy of recycling for profit (Cui & Forssberg, 2003). Other chemicals such as those used for insulation (plasticizers) or those used in solders such as lead are both toxic and less valuable and so are often handled in a less careful manner and by persons with little or no training such as children (Heacock et al., 2016; Perkins, Brune Drisse, Nxele, & Sly, 2014).

1.2 Quantities of Outdated Electronic Devices

The sheer quantities of outdated electronic devices being discarded, entering the waste stream or entering recycling programs each year, are in the millions of tons (UNEP, 2009). Many of these devices are dismantled and their components are safely recycled, but others are placed on container ships destined for developing countries where they are broken up and the valuable components

1

are recovered and the remaining waste is burned or buried in landfills (Robinson, 2009). Some of these devices may also be refurbished and have extended working lifetimes (Bovea, Ibanez-Fores, Perez-Belis, & Quemades-Beltran, 2016) as discussed below.

1.2.1 Number of Devices per Year

As noted above, the number of electronic devices entering the e-waste stream each year is in the hundreds of millions across a wide array of electronic products containing both valuable and toxic materials (ILO, 2012; Widmer, Oswald-Krapf, Sinha-Khetriwal, Schnellmann, & Böni, 2005). It has been estimated that ~80% of the devices submitted for recycling from developed countries are sent to developing countries both legally and illegally (ILO, 2012)

1.2.2 Tonnage per Year

The tonnage per year of discarded electronic products is on the order of millions of tons and growing at an exponential rate (UNEP, 2009). These materials now constitute the fastest growing segment of many municipal waste streams (ILO, 2012). If not efficiently recycled, these materials would contribute to solid waste problems in many countries due to leaching of toxic chemicals into soils and crops (Luo et al., 2011) and rivers and fish (Luo, Cai, & Wong, 2007).

2. REFURBISHING DISCARDED ELECTRONIC DEVICES

Another approach to recycling older electronic devices is to refurbish them via "reverse logistics" so that they will have extended working lifetimes by adding faster electronic circuits, RAM, and wireless capabilities (Ravi, 2012; White, Masanet, Rosen, & Beckman, 2003). These upgrades do reduce the levels of electronic waste, but the problem of what to do with older devices that are beyond repair and the old circuit boards or wiring, which is destined for destruction to recover metallic components, remains.

3. RECYCLING OF DEVICES MANUFACTURED WITH NEWER HIGH TECHNOLOGY ALLOY NANOMATERIALS

As noted above, e-waste is composed of a number of high technology materials from a rapidly evolving industry that introduces new materials in terms of both chemical and physical characteristics (e.g., nanomaterials) on an ongoing basis. This situation creates potential problems from the perspective of chemical safety and environmental dispersion (Bystrzejewska-Piotrowska, Golimowski, & Urban, 2009; Wynne, Buckley, Coumbe, Phillips, & Stevenson, 2008) since the database for these materials is usually very limited and especially so if device containing these chemicals are being recycled under non-regulated workspaces in developing countries by children as discussed below.

4. GLOBAL DISTRIBUTION STEAMS OF E-WASTE— WHERE DOES IT GO?

The global patterns of e-waste for recycling are well known and involve both developed countries (Kahat et al, 2008; Plambeck & Wong, 2009) such as the United States or European countries and developing countries such as Nigeria, Pakistan, Bangladesh, India, and China (Nnorom & Osibanjo, 2008; Robinson, 2009; Sthiannopkao & Wong, 2013; Streicher-Porte et al., 2005; UNEP, 2009; Widmer et al., 2005; C.S.C. Wong, Wu, Duzgoren-Aydin, 2007; M. Wong et al., 2007). The global pathways and major recycling sites are well known as shown in Fig. 1.1. The recycling of these devices has a number of both positive aspects (employment) and increased access to the Internet as examples, as well as some negative consequences such as toxic chemical pollution with increased disease burdens (Williams et al., 2008). Toxic materials from these activities such as metals have been reported to eventually find their way into groundwater (Chen, 2006) or sediments (C.S.C. Wong et al., 2007; M. Wong et al., 2007) in some Asian countries.

5. UPTAKE OF TOXIC CHEMICALS ORIGINATING FROM E-WASTE INTO FOOD

In many countries, water systems that may drain e-waste recycling areas are also used for irrigation of food crops and support commercial or sport fisheries, which are used as food sources. This means that toxic chemicals released during the recycling of e-waste materials may find their way into food webs that ultimately result in human exposures and attendant public health effects (Borthakur, 2016; Hibbert & Ogunseitan, 2014).

6. BIOLOGICAL EFFECTS OF E-WASTE CHEMICALS

Toxic chemicals released from e-waste materials may exert biological effects on environmental ecosystems including both plants and animals inhabiting aquatic and terrestrial ecosystems.

6.1 Ecosystems

Ecosystems are integrated diverse organization of plants and animals, which interact with each other usually on a number of levels. Healthy ecosystems are diverse and self-sustaining due to these interactions (Kirwan et al., 2009; Romero & Srivastava, 2010) but may become less so if environmental conditions reduce diversity to only those organisms that can function and reproduce under adverse conditions such as those produced by excess exposure to mixtures of toxic chemicals (Liu et al., 2008; Pramila, Fulekar, & Bhawana, 2012; Tang et al., 2010) such as those released by improper recycling of electronic

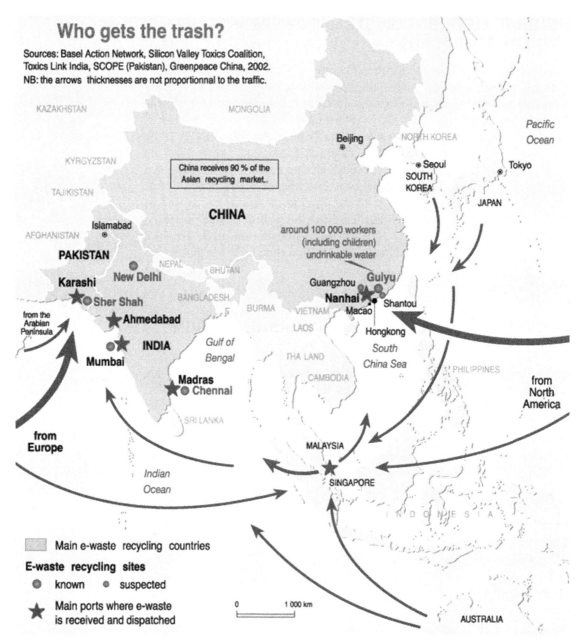

Who gets the trash?

Sources: Basel Action Network, Silicon Valley Toxics Coalition, Toxics Link India, SCOPE (Pakistan), Greenpeace China, 2002.
NB: the arrows thicknesses are not proportionnal to the traffic.

China receives 90 % of the Asian recycling market.

around 100 000 workers (including children) undrinkable water

Main e-waste recycling countries

E-waste recycling sites
● known   · suspected

★ Main ports where e-waste is received and dispatched

0   1 000 km

FIGURE 1.1
Global dispersion patterns of e-waste from developed to developing countries. *Basel Action Network, Silicon Valley Toxics Coalition, Toxics Link India, SCOPE (Pakistan), Green peace China, 2002.*

waste. This section will review some of the documented effects of e-waste chemicals on aquatic and terrestrial ecosystems and their potential relationships to human health effects.

6.1.1 Aquatic Ecosystems
Toxic chemicals released by industrial activities including improper recycling of electronic products may find their way into rivers, lakes, and oceans by atmospheric deposition, runoff, and leaching from corrosion of fabricated devices (Huang, Nkrumah, Anim, & Mensah, 2014). Among the known consequences of chemical exposures to a variety of chemicals are death of aquatic organisms (Malaj et al., 2014), reproductive failures from endocrine-disrupting chemicals, cancers (Lerebours et al., 2014), and altered susceptibility to infections from alterations of immune systems (Xu, Yang, Qiu, Pan, & Wu, 2013).

6.1.2 Terrestrial Ecosystems
Similar effects from such chemical exposures may occur in terrestrial ecosystems and influence crop plants (Luo et al., 2011; Wang, Luo, Li, Yin, & Zhang, 2014), birds (Luo et al., 2008; Zhang et al., 2015), including eggs (Zheng et al., 2012), and edible tissues of chickens (Liang et al., 2008), indicating disruptions of the food chain used to feed human populations or resulting in chemical contaminations of food such that it may become a health hazard following consumption. Lesser effects have been reported on rice paddy soil microbiological communities (Tang et al., 2014). Possible changes in the structure of microbial communities have also been reported (Zhang et al., 2010) near open-burning e-waste sites, so further research is needed into this important area in relation to crop production.

6.2 Human Health Effects
The human health effects of exposures to toxic chemicals derived from e-waste have also been reported in epidemiological studies among workers dealing with e-waste materials (Akormedi, Asampong, & Fobil, 2013; Asampong et al., 2015; Asante et al., 2012; Burns, Sun, Fobil, & Neitzel, 2016; Song & Li, 2014; Yang et al., 2013; Yuan et al., 2008) and the general population living near e-waste recycling sites (Ha et al., 2009; Tao et al., 2015). The general types of health effects reported include respiratory problems, skin lesions, reproductive disorders (Zeng, Xu, Boezen, & Huo, 2016; Zhou, Ju, Wu, & Yang, 2013), metabolic disorders, and cancers (Xu et al., 2012). In many of these situations, women and children bear a greater proportion of the adverse health burden (McAllister, Magee, & Hale, 2014; Perkins et al., 2014). In most situations, exposures are to mixtures of inorganic (lead, cadmium, arsenic, gallium, and indium) and organic (solvents, fire retardants, and plasticizers) chemicals, which may interact with each other in a variety of known and unknown ways (Wang et al., 2015). These complex exposures

complicate risk assessment strategies, which are further compounded by issues of child labor and nutritional deficiencies in developing countries (Heacock et al., 2016; Perkins et al., 2014; Shendell, Noomnual, Chishti, Sorensen Allacci, & Madrigano, 2016) and differences in gender susceptibility to chemical toxicity (Fowler et al., 2005, 2008). In addition, the immune-suppressive effects of many chemicals may add to the increased susceptibility of persons in developing countries to infectious diseases. The point here is that chemical exposures derived from unsafe e-waste recycling practices may add to existing public health disease burdens in developing countries where resources are already stretched.

7. REFURBISHING OF OUTDATED ELECTRONIC DEVICES

In addition to the practices focused on the destruction of outdated electronic devices for the purpose of recovering valuable metals such as gold, copper, gallium, and indium, electronic devices may be refurbished by interchanging parts such as circuit boards to extend the working life of a given device. Such devices may be resold for a profit. This is particularly true in developing countries where wages are generally low and child labor laws are limited if they exist at all and worker safety protections particularly regarding the use of lead-based solders to reconstitute a device are minimal if existing (Huo et al., 2007; Zheng et al., 2008). There are hence both an occupational safety and child labor aspects to the unregulated refurbishing of outdated electronic devices in developing countries. These issues will be discussed in greater detail in following sections.

8. INTERCOUNTRY VARIATIONS IN THE COLLECTION OF ELECTRONIC DEVICES FOR RECYCLING

The e-waste stream starts with the collection of discarded or outdated electronic devices, which shows a great deal of variability between countries in terms of degree of organization, regulatory oversight, and the ultimate destination of these devices for the physical dismantling and recycling activities. Most developed countries have organized recycling centers for the collection of electronic devices and distribution to licensed recyclers (Hicks, Dietmar, & Eugster, 2005; Sinha-Khetriwal, Kraeuchi, & Schwaninger, 2005). These formal organized operations work with varying degrees of effectiveness, and not all discarded electronics are captured in such centers and are passed on to informal recycling pathways (Miller, Duan, Gregory, Kahhat, & Kirchain, 2016). A significant fraction of discarded electronics enter into unregulated informal (Widmer et al., 2005) operations usually operated by groups or families with members in both the developed and developing countries (Chi, Streicher-Porte, Wang, & Reuter, 2011) who utilized container shipping to move discarded electronics between countries on a large scale. It is interesting to note that these informal networks are both well organized and well known with regard to destinations (UNEP, 2009). E-waste from European countries is largely shipped to countries such as Pakistan, Bangladesh, and India (Streicher-Porte et al., 2005), while electronics from the United States and Canada are largely destined for China and cities such as Guiyang in particular (UNEP, 2009) where disruption of hormonal homeostasis and increased rates of adverse birth outcomes have been reported in childbearing women (Xu et al., 2012; Zhou et al., 2013).

9. RECYCLING OF COMPONENT MATERIALS IN ELECTRONIC DEVICES

Electronic devices are composed of multiple components that are usually separated during the initial dismantling of the devices and entered into different waste streams depending on their potential value. For example, circuit boards and computer hard drives containing valuable metals such as gold, indium, and gallium are carefully handled prior to recycling (Cayumil, Khanna, Rajarao, Mukherjee, & Sahajwalla, 2015; Fowler & Maples-Reynolds, 2015; Fowler & Sexton, 2002, 2015). Insulated copper wiring is stripped off the plasticized insulation to recover the copper metal, but this process may not be conducted under the safe working conditions in developing countries (Ni & Zeng, 2009). Other less valuable metals such as lead from solders or steel alloys used in refrigerator or washing machine parts may be simply dumped in landfills. Plastic components may be recovered or dumped depending on the country and availability of reprocessing facilities for handling the chemicals in plastics (Hu et al., 2013; Muenhor, Harrad, Ali, & Covaci, 2010; Wang et al., 2013; Wang & Xu, 2014). The bottom line here is that the recycling of components from electronic devices is both complex and evolving. The occupational and environmental safety aspects of this overall process are highly variable and dependent on the countries in which the recycling activities are being conducted (Liu, Tanaka, & Matsui, 2006; Nnorom & Osibanjo, 2008). Developing countries are at particular risk for adverse consequences due to a general lack of regulatory oversight.

10. DIFFERENCES IN E-WASTE HANDLING BETWEEN DEVELOPED AND DEVELOPING COUNTRIES

As noted above, there are marked differences between countries in the handling of electronic waste and the degree with which these activities are regulated. These differences are particularly striking between developed countries, which are the primary points of origin for e-waste streams, and the developing countries or regions of the world where artisanal recycling operations are most commonly located (Sthiannopkao & Wong, 2013). On the one hand, the recycling of valuable materials such as gold, indium, gallium, and copper is positive in terms of both reducing the volume of materials sent to landfills and municipal incinerators (Li, Yu, Sheng, Fu, & Peng, 2007) and providing income for persons with limited access to education and job opportunities in developing countries (Buchmann & Hannum, 2001). On the other hand, the lack of occupational and environmental oversight in many developing countries has a number of adverse consequences including (1) increased rates of infectious diseases due to interactions between chemicals and the immune system (Frazzoli, Orisakwe, Dragone, & Mantovani, 2010) and (2) environmental damage due to disposal of unrecycled components into land areas (C.S.C. Wong et al., 2007; M. Wong et al., 2007), rivers, and estuaries (Luo et al., 2007) utilized for growing crops (Zhang et al., 2014) and harvesting fishery products (Wu et al., 2008). The short-term and long-term net consequences of these anabolic and catabolic forces for the transition of developing countries into more self-sustaining societies are still not clear.

11. CHILD LABOR

As noted above, the shipment of e-waste materials to developing countries includes the use of child labor under unregulated conditions to recover valuable metals from computer hard drives and circuit boards and wiring from computers, cell phones, printers, and televisions. These activities result in the exposure of children to chemicals given off during destruction of the solid components such as particles and vapor phase materials released during burning of plasticized insulation around copper wiring. Since organ systems of children are growing and developing, these exposures may have long-term health consequences later in life (Heacock et al., 2016).

12. OCCUPATIONAL AND ENVIRONMENTAL SAFETY ISSUES

Many developing countries have few, if any, occupational or environmental laws governing the handling of toxic materials including those released during the recycling of electronic devices. Open-pit burning of wiring to remove insulation is one issue with both occupational exposures for the workers tending the fires and environmental issues with regard to atmospheric release of volatile chemicals, which may impact local communities near the burn sites (C.S.C. Wong et al., 2007; M. Wong et al., 2007).

13. LANDFILL OPERATIONS

As noted above, not all of the materials in e-waste can be recovered, and there will always be some residual materials that will remain. In many countries, these materials will be simply buried in landfills, which is true in both developed and developing countries (Greenpeace, 2009). Potential environmental and health problems may occur if these land areas are also used for agriculture (Shen et al., 2009) or produce runoff to local streams, lakes, and shallow water embayments used for production of fishery products, resulting in contamination of edible seafood (Xing, Chan, Leung, Wu, & Wong, 2009). Toxic chemicals released from degradation of buried or dumped e-waste materials over time may find their way into the food chain of livestock or humans (Fu et al., 2008). This is particularly true for countries where there are food shortages and all available avenues of food production must be utilized (Pretty, Morison, & Hine, 2003).

14. WASTE PONDS

Disposal of e-waste components into water impoundments is another known approach for disposal of industrial wastes in many countries (Leung, Cai, & Wong, 2006), and this approach is undoubtedly utilized for e-waste components that cannot be recycled for economic reasons. Again, the issue of release of toxic chemicals into water systems used for drinking water and food production is of concern if the impoundments are not secure. There is also the known breakdown of earthen impoundment walls resulting in the massive spillage of large quantities of contaminated water into adjacent streams and rivers resulting in fish kills and long-term contamination of system.

15. INCINERATION

Incineration of unrecycled e-waste is a major concern for municipal incineration operations since small particulate and vapor phase aerosols may be dispersed over large areas resulting in local, national, and international human chemical exposures (Greenpeace, 2009; Ma et al., 2009). The atmospheric concentrations of chemicals related to e-waste have been reported to show diurnal variation in two areas of China (Chen et al., 2009).

References

Akormedi, M., Asampong, E., & Fobil, J. N. (2013). Working conditions and environmental exposures among electronic waste workers in Ghana. *International Journal of Occupational and Environmental Health*, 19(4), 278–286. http://dx.doi.org/10.1179/2049396713y.0000000034.

Asampong, E., Dwuma-Badu, K., Stephens, J., Srigboh, R., Neitzel, R., Basu, N., & Fobil, J. N. (2015). Health seeking behaviours among electronic waste workers in Ghana. *BMC Public Health*, 15, 1065. http://dx.doi.org/10.1186/s12889-015-2376-z.

Asante, K. A., Agusa, T., Biney, C. A., Agyekum, W. A., Bello, M., Otsuka, M., … Tanabe, S. (2012). Multi-trace element levels and arsenic speciation in urine of e-waste recycling workers from Agbogbloshie, Accra in Ghana. *The Science of the Total Environment*, 424, 63–73. http://dx.doi.org/10.1016/j.scitotenv.2012.02.072.

Borthakur, A. (2016). Health and environmental hazards of electronic waste in India. *Journal of Environmental Health, 78*(8), 18–23.

Bovea, M. D., Ibanez-Fores, V., Perez-Belis, V., & Quemades-Beltran, P. (2016). Potential reuse of small household waste electrical and electronic equipment: Methodology and case study. *Waste Management, 53*, 204–217. http://dx.doi.org/10.1016/j.wasman.2016.03.038.

Buchmann, C., & Hannum, E. (2001). Education and stratification in developing countries: A review of theories and research. *Annual Review of Sociology*, 77–102.

Burns, K. N., Sun, K., Fobil, J. N., & Neitzel, R. L. (2016). Heart rate, stress, and occupational noise exposure among electronic waste recycling workers. *International Journal of Environmental Research and Public Health, 13*(1). http://dx.doi.org/10.3390/ijerph13010140.

Bystrzejewska-Piotrowska, G., Golimowski, J., & Urban, P. L. (2009). Nanoparticles: Their potential toxicity, waste and environmental management. *Waste Management, 29*(9), 2587–2595. http://dx.doi.org/10.1016/j.wasman.2009.04.001.

Cayumil, R., Khanna, R., Rajarao, R., Mukherjee, P. S., & Sahajwalla, V. (2015). Concentration of precious metals during their recovery from electronic waste. *Waste Management*. http://dx.doi.org/10.1016/j.wasman.2015.12.004.

Chen, D., Bi, X., Zhao, J., Chen, L., Tan, J., Mai, B., … Wong, M. (2009). Pollution characterization and diurnal variation of PBDEs in the atmosphere of an e-waste dismantling region. *Environmental Pollution, 157*(3), 1051–1057. http://dx.doi.org/10.1016/j.envpol.2008.06.005.

Chen, H.-W. (2006). Gallium, indium and arsenic pollution of groundwater from a semiconductor manufacturing area of Taiwan. *Bulletin of Environmental Contamination and Toxicology, 77*, 289–296.

Chi, X., Streicher-Porte, M., Wang, M. Y. L., & Reuter, M. A. (2011). Informal electronic waste recycling: A sector review with special focus on China. *Waste Management, 31*(4), 731–742. http://dx.doi.org/10.1016/j.wasman.2010.11.006.

Cui, J., & Forssberg, E. (2003). Mechanical recycling of waste electric and electronic equipment: A review. *Journal of Hazardous Materials, 99*(3), 243–263.

Fowler, B. A., Conner, E. A., & Yamauchi, H. (2005). Metabolomic and proteomic biomarkers for III-V semiconductors: Chemical-specific porphyrinurias and proteinurias. *Toxicology and Applied Pharmacology, 206*(2), 121–130.

Fowler, B. A., Conner, E. A., & Yamauchi, H. (2008). Proteomic and metabolomic biomarkers for III-V semiconductors: Prospects for applications to nano-materials. *Toxicology and Applied Pharmacology, 233*(1), 110–115.

Fowler, B. A., & Maples-Reynolds, N. (2015). Indium. In G. F. Nordberg, B. A. Fowler, & M. Nordberg (Eds.), *Handbook on the toxicology of metals* (4th ed.) (pp. 845–853). Amsterdam: Elsevier Publishers.

Fowler, B. A., & Sexton, M. J. (2002). Semiconductor metals. In B. Sarkar (Ed.), *Heavy metals in the environment* (pp. 631–646). New York: Marcel Dekker Publishers.

Fowler, B. A., & Sexton, M. J. (2015). Gallium and semiconductor compounds. In G. F. Nordberg, B. A. Fowler, & M. Nordberg (Eds.), *Handbook on the toxicology of metals* (4th ed.) (pp. 787–797). Amsterdam: Elsevier Publishers.

Frazzoli, C., Orisakwe, O. E., Dragone, R., & Mantovani, A. (2010). Diagnostic health risk assessment of electronic waste on the general population in developing countries' scenarios. *Environmental Impact Assessment Review, 30*(6), 388–399.

Fu, J., Zhou, Q., Liu, J., Liu, W., Wang, T., Zhang, Q., & Jiang, G. (2008). High levels of heavy metals in rice (*Oryza sativa* L.) from a typical E-waste recycling area in southeast China and its potential risk to human health. *Chemosphere, 71*(7), 1269–1275.

Greenpeace. (2009). *Where does e-waste end up*. www.greenpeace.org.

Ha, N. N., Agusa, T., Ramu, K., Tu, N. P., Murata, S., Bulbule, K. A., … Tanabe, S. (2009). Contamination by trace elements at e-waste recycling sites in Bangalore, India. *Chemosphere, 76*(1), 9–15. http://dx.doi.org/10.1016/j.chemosphere.2009.02.056.

Hagiya, M., Konagaya, A., Kobayashi, S., Saito, H., & Murata, S. (2014). Molecular robots with sensors and intelligence. *Accounts of Chemical Research, 47*(6), 1681–1690. http://dx.doi.org/10.1021/ar400318d.

Heacock, M., Kelly, C. B., Asante, K. A., Birnbaum, L. S., Bergman, A. L., Brune, M. N., … Suk, W. A. (2016). E-waste and harm to vulnerable populations: A growing global problem. *Environmental Health Perspectives, 124*(5), 550–555. http://dx.doi.org/10.1289/ehp.1509699.

Hibbert, K., & Ogunseitan, O. A. (2014). Risks of toxic ash from artisanal mining of discarded cellphones. *Journal of Hazardous Materials, 278*, 1–7. http://dx.doi.org/10.1016/j.jhazmat.2014.05.089.

Hicks, C., Dietmar, R., & Eugster, M. (2005). The recycling and disposal of electrical and electronic waste in China—legislative and market responses. *Environmental Impact Assessment Review, 25*(5), 459–471. http://dx.doi.org/10.1016/j.eiar.2005.04.007.

Hu, J., Xiao, X., Peng, P., Huang, W., Chen, D., & Cai, Y. (2013). Spatial distribution of polychlorinated dibenzo-p-dioxins and dibenzo-furans (PCDDs/Fs) in dust, soil, sediment and health risk assessment from an intensive electronic waste recycling site in Southern China. *Environmental Science Processes and Impacts, 15*(10), 1889–1896. http://dx.doi.org/10.1039/c3em00319a.

Huang, J., Nkrumah, P. N., Anim, D. O., & Mensah, E. (2014). E-waste disposal effects on the aquatic environment: Accra, Ghana. *Reviews of Environmental Contamination and Toxicology, 229*, 19–34. http://dx.doi.org/10.1007/978-3-319-03777-6_2.

Huo, X., Peng, L., Xu, X., Zheng, L., Qiu, B., Qi, Z., … Piao, Z. (2007). Elevated blood lead levels of children in Guiyu, an electronic waste recycling town in China. *Environmental Health Perspectives*, 1113–1117.

ILO. (2012). *The global impact of e-waste: Addressing the challenge*. Geneva, 71 pp.

Kahat, R., Kim, J., Xu, M., Allenby, B., Williams, E., & Zhang, P. (2008). Exploring e-waste management systems in the United States. *Resources Conservation and Recycling, 52*(7), 955–964.

Kirwan, L., Connolly, J., Finn, J., Brophy, C., Lüscher, A., Nyfeler, D., & Sebastia, M. (2009). Diversity–interaction modeling: Estimating contributions of species identities and interactions to ecosystem function. *Ecology, 90*(8), 2032–2038.

Lerebours, A., Stentiford, G. D., Lyons, B. P., Bignell, J. P., Derocles, S. A., & Rotchell, J. M. (2014). Genetic alterations and cancer formation in a European flatfish at sites of different contaminant burdens. *Environmental Science and Technology, 48*(17), 10448–10455. http://dx.doi.org/10.1021/es502591p.

Leung, A., Cai, W. Z., & Wong, H. M. (2006). Environmental contamination from electronic waste recycling at Guiyu, southeast China. *Journal of Material Cycles and Waste Management, 8*(1), 21–33. http://dx.doi.org/10.1007/s10163-005-0141-6.

Li, H., Yu, L., Sheng, G., Fu, J., & Peng, P. (2007). Severe PCDD/F and PBDD/F pollution in air around an electronic waste dismantling area in China. *Environmental Science and Technology, 41*(16), 5641–5646.

Liang, S. X., Zhao, Q., Qin, Z. F., Zhao, X. R., Yang, Z. Z., & Xu, X. B. (2008). Levels and distribution of polybrominated diphenyl ethers in various tissues of foraging hens from an electronic waste recycling area in South China. *Environmental Toxicology and Chemistry, 27*(6), 1279–1283. http://dx.doi.org/10.1897/07-518.1.

Liu, H., Zhou, Q., Wang, Y., Zhang, Q., Cai, Z., & Jiang, G. (2008). E-waste recycling induced polybrominated diphenyl ethers, polychlorinated biphenyls, polychlorinated dibenzo-p-dioxins and dibenzo-furans pollution in the ambient environment. *Environment International, 34*(1), 67–72. http://dx.doi.org/10.1016/j.envint.2007.07.008.

Liu, X., Tanaka, M., & Matsui, Y. (2006). Electrical and electronic waste management in China: Progress and the barriers to overcome. *Waste Management and Research: The Journal of the International Solid Wastes and Public Cleansing Association, ISWA, 24*(1), 92–101.

Luo, C., Liu, C., Wang, Y., Liu, X., Li, F., Zhang, G., & Li, X. (2011). Heavy metal contamination in soils and vegetables near an e-waste processing site, south China. *Journal of Hazardous Materials, 186*(1), 481–490. http://dx.doi.org/10.1016/j.jhazmat.2010.11.024.

Luo, Q., Cai, Z. W., & Wong, M. H. (2007). Polybrominated diphenyl ethers in fish and sediment from river polluted by electronic waste. *Science of the Total Environment, 383*(1), 115–127.

Luo, X.-J., Zhang, X.-L., Liu, J., Wu, J.-P., Luo, Y., Chen, S.-J., … Yang, Z.-Y. (2008). Persistent halogenated compounds in waterbirds from an e-waste recycling region in South China. *Environmental Science and Technology, 43*(2), 306–311.

Ma, J., Horii, Y., Cheng, J., Wang, W., Wu, Q., Ohura, T., & Kannan, K. (2009). Chlorinated and parent polycyclic aromatic hydrocarbons in environmental samples from an electronic waste recycling facility and a chemical industrial complex in China. *Environmental Science and Technology, 43*(3), 643–649.

Malaj, E., von der Ohe, P. C., Grote, M., Kuhne, R., Mondy, C. P., Usseglio-Polatera, P., … Schafer, R. B. (2014). Organic chemicals jeopardize the health of freshwater ecosystems on the continental scale. *Proceedings of the National Academy of Sciences of the United States of America, 111*(26), 9549–9554. http://dx.doi.org/10.1073/pnas.1321082111.

McAllister, L., Magee, A., & Hale, B. (2014). Women, e-waste, and technological solutions to climate change. *Health and Human Rights, 16*(1), 166–178.

Miller, T. R., Duan, H., Gregory, J., Kahhat, R., & Kirchain, R. (2016). Quantifying domestic used electronics flows using a combination of material flow methodologies: A US case study. *Environmental Science and Technology, 50*(11), 5711–5719. http://dx.doi.org/10.1021/acs.est.6b00079.

Muenhor, D., Harrad, S., Ali, N., & Covaci, A. (2010). Brominated flame retardants (BFRs) in air and dust from electronic waste storage facilities in Thailand. *Environment International, 36*(7), 690–698. http://dx.doi.org/10.1016/j.envint.2010.05.002.

Ni, H.-G., & Zeng, E. Y. (2009). Law enforcement and global collaboration are the keys to containing e-waste tsunami in China. *Environmental Science and Technology, 43*(11), 3991–3994.

Nnorom, I. C., & Osibanjo, O. (2008). Electronic waste (e-waste): Material flows and management practices in Nigeria. *Waste Management, 28*(8), 1472–1479. http://dx.doi.org/10.1016/j.wasman.2007.06.012.

Perkins, D. N., Brune Drisse, M. N., Nxele, T., & Sly, P. D. (2014). E-waste: A global hazard. *Annals of Global Health, 80*(4), 286–295. http://dx.doi.org/10.1016/j.aogh.2014.10.001.

Plambeck, E., & Wong, O. (2009). Effects of e-waste regulation on new products introduction. *Management Science, XX*, 333–349.

Pramila, S., Fulekar, M., & Bhawana, P. (2012). E-waste-a challenge for tomorrow. *Research Journal of Recent Sciences.* ISSN: 2277-2502.

Pretty, J. N., Morison, J. I., & Hine, R. E. (2003). Reducing food poverty by increasing agricultural sustainability in developing countries. *Agriculture, Ecosystems and Environment, 95*(1), 217–234.

Ravi, V. (2012). Selection of third-party reverse logistics providers for End-of-Life computers using TOPSIS-AHP based approach. *International Journal of Logistics Systems and Management, 11*(1), 24–37. http://dx.doi.org/10.1504/IJLSM.2012.044048.

Robinson, B. H. (2009). E-waste: An assessment of global production and environmental impacts. *Science of the Total Environment, 408*(2), 183–191. http://dx.doi.org/10.1016/j.scitotenv.2009.09.044.

Romero, G. Q., & Srivastava, D. S. (2010). Food-web composition affects cross-ecosystem interactions and subsidies. *Journal of Animal Ecology, 79*(5), 1122–1131.

Saphores, J. D., Nixon, H., Ogunseitan, O. A., & Shapiro, A. A. (2009). How much e-waste is there in US basements and attics? Results from a national survey. *Journal of Environmental Management, 90*(11), 3322–3331. http://dx.doi.org/10.1016/j.jenvman.2009.05.008.

Shen, C., Chen, Y., Huang, S., Wang, Z., Yu, C., Qiao, M., … Zhu, Y. (2009). Dioxin-like compounds in agricultural soils near e-waste recycling sites from Taizhou area, China: Chemical and bioanalytical characterization. *Environment International, 35*(1), 50–55.

Shendell, D. G., Noomnual, S., Chishti, S., Sorensen Allacci, M., & Madrigano, J. (2016). Exposures resulting in safety and health concerns for child laborers in less developed countries. *Journal of Environmental and Public Health, 2016*, 3985498. http://dx.doi.org/10.1155/2016/3985498.

Singh, N., Li, J., & Zeng, X. (2016). Global responses for recycling waste CRTs in e-waste. *Waste Management, 5.*

Sinha-Khetriwal, D., Kraeuchi, P., & Schwaninger, M. (2005). A comparison of electronic waste recycling in Switzerland and in India. *Environmental Impact Assessment Review, 25*(5), 492–504. http://dx.doi.org/10.1016/j.eiar.2005.04.006.

Song, Q., & Li, J. (2014). Environmental effects of heavy metals derived from the e-waste recycling activities in China: A systematic review. *Waste Management, 34*(12), 2587–2594. http://dx.doi.org/10.1016/j.wasman.2014.08.012.

Sthiannopkao, S., & Wong, M. H. (2013). Handling e-waste in developed and developing countries: Initiatives, practices, and consequences. *The Science of the Total Environment, 463–464*, 1147–1153. http://dx.doi.org/10.1016/j.scitotenv.2012.06.088.

Streicher-Porte, M., Widmer, R., Jain, A., Bader, H.-P., Scheidegger, R., & Kytzia, S. (2005). Key drivers of the e-waste recycling system: Assessing and modelling e-waste processing in the informal sector in Delhi. *Environmental Impact Assessment Review, 25*(5), 472–491. http://dx.doi.org/10.1016/j.eiar.2005.04.004.

Tang, X., Hashmi, M. Z., Long, D., Chen, L., Khan, M. I., & Shen, C. (2014). Influence of heavy metals and PCBs pollution on the enzyme activity and microbial community of paddy soils around an e-waste recycling workshop. *International Journal of Environmental Research and Public Health, 11*(3), 3118–3131. http://dx.doi.org/10.3390/ijerph110303118.

Tang, X., Shen, C., Shi, D., Cheema, S. A., Khan, M. I., Zhang, C., & Chen, Y. (2010). Heavy metal and persistent organic compound contamination in soil from Wenling: An emerging e-waste recycling city in Taizhou area, China. *Journal of Hazardous Materials, 173*(1–3), 653–660. http://dx.doi.org/10.1016/j.jhazmat.2009.08.134.

Tao, X. Q., Shen, D. S., Shentu, J. L., Long, Y. Y., Feng, Y. J., & Shen, C. C. (2015). Bioaccessibility and health risk of heavy metals in ash from the incineration of different e-waste residues. *Environmental Science and Pollution Research International, 22*(5), 3558–3569. http://dx.doi.org/10.1007/s11356-014-3562-8.

UNEP. (2009). *Recycling from e-waste resources (StEP) sustainable innovation and technology transfer industrial sector studies.*

Wang, J., Liu, L., Wang, J., Pan, B., Fu, X., Zhang, G., … Lin, K. (2015). Distribution of metals and brominated flame retardants (BFRs) in sediments, soils and plants from an informal e-waste dismantling site, South China. *Environmental Science and Pollution Research International, 22*(2), 1020–1033. http://dx.doi.org/10.1007/s11356-014-3399-1.

Wang, P., Zhang, H., Fu, J., Li, Y., Wang, T., Wang, Y., … Jiang, G. (2013). Temporal trends of PCBs, PCDD/Fs and PBDEs in soils from an E-waste dismantling area in East China. *Environmental Science. Processes and Impacts, 15*(10), 1897–1903. http://dx.doi.org/10.1039/c3em00297g.

Wang, R., & Xu, Z. (2014). Recycling of non-metallic fractions from waste electrical and electronic equipment (WEEE): A review. *Waste Management, 34*(8), 1455–1469. http://dx.doi.org/10.1016/j.wasman.2014.03.004.

Wang, Y., Luo, C., Li, J., Yin, H., & Zhang, G. (2014). Influence of plants on the distribution and composition of PBDEs in soils of an e-waste dismantling area: Evidence of the effect of the rhizosphere and selective bioaccumulation. *Environmental Pollution, 186,* 104–109. http://dx.doi.org/10.1016/j.envpol.2013.11.018.

White, C. D., Masanet, E., Rosen, C. M., & Beckman, S. L. (2003). Product recovery with some byte: An overview of management challenges and environmental consequences in reverse manufacturing for the computer industry. *Journal of Cleaner Production, 11*(4), 445–458. http://dx.doi.org/10.1016/S0959-6526(02)00066-5.

Widmer, R., Oswald-Krapf, H., Sinha-Khetriwal, D., Schnellmann, M., & Böni, H. (2005). Global perspectives on e-waste. *Environmental Impact Assessment Review, 25*(5), 436–458. http://dx.doi.org/10.1016/j.eiar.2005.04.001.

Wikipedia. (2016). http://en.Wikipedia.org/wiki.Electronic_waste.

Williams, E., Kahhat, R., Allenby, B., Kavazanjian, E., Kim, J., & Xu, M. (2008). Environmental, social, and economic implications of global reuse and recycling of personal computers. *Environmental Science and Technology, 42*(17), 6446–6454.

Wong, C. S. C., Wu, S. C., & Duzgoren-Aydin, N. S. (2007). Trace metal contamination of sediments in an e-waste processing village in China. *Environmental Pollution, 145*(2), 434–442.

Wong, M., Wu, S., Deng, W., Yu, X., Luo, Q., Leung, A., ... Wong, A. (2007). Export of toxic chemicals–a review of the case of uncontrolled electronic-waste recycling. *Environmental Pollution, 149*(2), 131–140.

Wu, J.-P., Luo, X.-J., Zhang, Y., Luo, Y., Chen, S.-J., Mai, B.-X., & Yang, Z.-Y. (2008). Bioaccumulation of polybrominated diphenyl ethers (PBDEs) and polychlorinated biphenyls (PCBs) in wild aquatic species from an electronic waste (e-waste) recycling site in South China. *Environment International, 34*(8), 1109–1113.

Wynne, J. H., Buckley, J. L., Coumbe, C. E., Phillips, J. P., & Stevenson, S. (2008). Reducing hazardous material and environmental impact through recycling of scandium nano-material waste. *Journal of Environmental Science and Health, Part A, 43*(4), 357–360. http://dx.doi.org/10.1080/10934520701795483.

Xing, G. H., Chan, J. K. Y., Leung, A. O. W., Wu, S. C., & Wong, M. H. (2009). Environmental impact and human exposure to PCBs in Guiyu, an electronic waste recycling site in China. *Environment International, 35*(1), 76–82. http://dx.doi.org/10.1016/j.envint.2008.07.025.

Xu, H., Yang, M., Qiu, W., Pan, C., & Wu, M. (2013). The impact of endocrine-disrupting chemicals on oxidative stress and innate immune response in zebrafish embryos. *Environmental Toxicology and Chemistry, 32*(8), 1793–1799. http://dx.doi.org/10.1002/etc.2245.

Xu, X., Yang, H., Chen, A., Zhou, Y., Wu, K., Liu, J., ... Huo, X. (2012). Birth outcomes related to informal e-waste recycling in Guiyu, China. *Reproductive Toxicology, 33*(1), 94–98. http://dx.doi.org/10.1016/j.reprotox.2011.12.006.

Yang, Q., Qiu, X., Li, R., Liu, S., Li, K., Wang, F., ... Zhu, T. (2013). Exposure to typical persistent organic pollutants from an electronic waste recycling site in Northern China. *Chemosphere, 91*(2), 205–211. http://dx.doi.org/10.1016/j.chemosphere.2012.12.051.

Yuan, J., Chen, L., Chen, D., Guo, H., Bi, X., Ju, Y., ... Chen, X. (2008). Elevated serum polybrominated diphenyl ethers and thyroid-stimulating hormone associated with lymphocytic micronuclei in Chinese workers from an E-waste dismantling site. *Environmental Science and Technology, 42*(6), 2195–2200.

Zeng, X., Xu, X., Boezen, H. M., & Huo, X. (2016). Children with health impairments by heavy metals in an e-waste recycling area. *Chemosphere, 148,* 408–415. http://dx.doi.org/10.1016/j.chemosphere.2015.10.078.

Zhang, Q., Wu, J., Sun, Y., Zhang, M., Mai, B., Mo, L., … Zou, F. (2015). Do bird assemblages predict susceptibility by e-waste pollution? A comparative study based on species- and guild-dependent responses in China agroecosystems. *PLoS One, 10*(3), e0122264. http://dx.doi.org/10.1371/journal.pone.0122264.

Zhang, S., Xu, X., Wu, Y., Ge, J., Li, W., & Huo, X. (2014). Polybrominated diphenyl ethers in residential and agricultural soils from an electronic waste polluted region in South China: Distribution, compositional profile, and sources. *Chemosphere, 102*, 55–60. http://dx.doi.org/10.1016/j.chemosphere.2013.12.020.

Zhang, W., Wang, H., Zhang, R., Yu, X. Z., Qian, P. Y., & Wong, M. H. (2010). Bacterial communities in PAH contaminated soils at an electronic-waste processing center in China. *Ecotoxicology, 19*(1), 96–104. http://dx.doi.org/10.1007/s10646-009-0393-3.

Zheng, L., Wu, K., Li, Y., Qi, Z., Han, D., Zhang, B., … Chen, S. (2008). Blood lead and cadmium levels and relevant factors among children from an e-waste recycling town in China. *Environmental Research, 108*(1), 15–20.

Zheng, X. B., Wu, J. P., Luo, X. J., Zeng, Y. H., She, Y. Z., & Mai, B. X. (2012). Halogenated flame retardants in home-produced eggs from an electronic waste recycling region in South China: Levels, composition profiles, and human dietary exposure assessment. *Environment International, 45*, 122–128. http://dx.doi.org/10.1016/j.envint.2012.04.006.

Zhou, X., Ju, Y., Wu, Z., & Yang, K. (2013). Disruption of sex hormones and oxidative homeostasis in parturient women and their matching fetuses at an e-waste recycling site in China. *International Journal of Occupational and Environmental Health, 19*(1), 22–28. http://dx.doi.org/10.1179/2049396712y.0000000017.

Further Reading

Fowler, B. A. (Ed.). (2013). *Computational toxicology: Applications for risk assessment* (p. 258). Amsterdam: Elsevier Publishers.

Fowler, B. A. (2016). *Molecular biological markers for toxicology and risk assessment*. Amsterdam: Elsevier Publishers. 153 pp.

Nordberg, G. F., Fowler, B. A., & Nordberg, M. (2015). *Handbook on the toxicology of metals*. Amsterdam: Elsevier Publishers, 1385.

Wang, Q., He, A. M., Gao, B., et al. (2011). Increased levels of lead in blood and frequencies of lymphocytic micronucleaded cells among workers in an electronic-waste recycling site. *Journal of Environmental Science and Health. Part A, Toxic Hazardous Substance and Environmental Engineering, 46*(6), 669–676.

CHAPTER 2

Metals, Metallic Compounds, Organic Chemicals, and E-Waste Chemical Mixtures

INTRODUCTION

1. METALS AND METALLIC COMPOUNDS

There are a number of metals and metallic compounds present in electronic devices including lead, cadmium, arsenic, mercury as well as gold, gallium, indium arsenic, selenium, and antimony (Leung, Duzgoren-Aydin, Cheung, & Wong, 2008; Wong, Duzgoren-Aydin, Aydin, & Wong, 2007; Zheng et al., 2013). Some of these elements are highly valuable and extensive efforts are expended to recover them by recyclers, but others such as lead, cadmium, arsenic, and mercury are less valuable and may not be recovered and released into the environment with the resultant human exposures in air, food, and water (Sepúlveda et al., 2010; Wong, Duzgoren-Aydin, et al., 2007). These exposures may hence occur as mixtures of these elements (Cui & Zhang, 2008; Robinson, 2009). The types of effects resulting from mixtures of metallic compounds have been studied in both in vivo (Conner, Yamauchi, & Fowler, 1995; Goering, Maronpot, & Fowler, 1988; Mahaffey, Capar, Gladen, & Fowler, 1981; Whittaker et al., 2011) and in vitro (Aoki et al., 1990; Bustamente, Dock, Vahter, Fowler, & Orrenius, 1997; Fowler, Conne, & Yamauchi, 2005, 2008; Madden & Fowler, 2000, 2002) test systems. In addition, it is important to note that e-waste materials also contain plastics and a number of toxic organic chemical compounds (Robinson, 2009; Wong, Wu, et al., 2007), and possible interactions between metallic and organic constituents of e-waste must also be considered in any risk assessment approach for populations exposed to chemicals released from e-waste. The increasing use of nanomaterials in the fabrication of electronic devices (Caballero-Guzman, Sun, & Nowack, 2015) adds another level of complexity to any risk assessment for populations exposed to e-waste chemicals during recycling processes or released as a result of disposal of electronic devices into landfills or water bodies used for drinking or production of fishery products for human consumption. A more extensive discussion of the mechanisms of toxicity at the cellular level from exposure to a number of these chemical entities on an individual or mixture basis is provided below.

2. NANOMATERIALS

In the quest to produce smaller, lighter, and faster electronic devices, electronic manufacturers have moved increasingly to nanomaterial-based semiconductors such as particles made of indium arsenide, indium phosphide, and cadmium selenide (Heeres et al., 2007; Landi et al., 2005; Mushonga, Onani, Madiehe, & Meyer, 2012). In addition to metallic nanomaterials, organic nanomaterials containing chemicals derived from plastics (Zhuo & Levendis, 2014) should also be evaluated from a chemical safety perspective. While the behavior of nanomaterials released into the environment has received some attention to date (Klaine et al., 2008; Lowry, Gregory, Apte, & Lead, 2012; Vejerano, Leon, Holder, & Marr, 2014; Walser et al., 2012), further research is clearly needed to assess potential health effects of nanomaterials released into the environment in relation to open air recycling of this new generation of electronic devices containing these materials. Monitoring flows of both metallic and organic nanomaterials through the recycling process (Caballero-Guzman et al., 2015) is clearly an excellent idea from the perspective of both occupational and environmental risk assessment. This is an important and still unresolved area of public health research since it incorporates multimedia exposures, populations at special risk such as children in relation to occupational exposures, and, potentially, dispersion of nanoparticles over large areas.

3. REPRESENTATIVE ORGANIC E-WASTE CHEMICALS

As noted above, e-waste is chemically composed of both inorganic and organic chemicals, which each have their own toxic properties on both an individual or mixture basis. These potential interactive effects are further complicated by the issue of child labor involvement in recycling activities in developing countries. The exposure of children to e-waste chemicals during development may lead to latent health effects such as diabetes or cancer later in life (Heindel, 2003; Jirtle & Skinner, 2007; Perera & Herbstman, 2011; Skinner, Manikkam, & Guerrero-Bosagna, 2011). Common chemicals such as bisphenol A (BPA) (Huang, Zhao, Liu, & Sun, 2014; Matsukami et al., 2015), PBBs, PCBs, PBDEs (Zhao et al., 2009), and styrene (Kiran, Ekinci, & Snape, 2000) are well-known toxic chemicals that are used in the production of electronic devices. Human exposures may occur to these chemicals during recycling via open-pit burning of wiring and circuit boards and incineration of plastic composite computer housings. The overall point is that open-pit incineration of old electronic devices will release a number of toxic chemicals to which persons may be exposed. These chemicals have been shown to include a number of PAHs and PCBs as well as a number of the toxic metals noted above (Tang et al., 2014). In addition, a number of other persistent organic pollutants such as PBDEs, PBBs, dibenzo-dioxins, and dibenzofurans tetrabromo BPA compounds (Ni, Zeng, Tao, & Zeng, 2010; Shen et al., 2009) may be released from electronic devices during recycling activities. A number of these chemicals have been linked to obesity (Heindel & vom Saal, 2009) and metabolic diseases such as type II diabetes (Heindel & vom Saal, 2009), which may occur via disruption of endocrine regulatory pathways (Heindel & vom Saal, 2009). The main point here is that there are a large number of toxic organic chemicals present in these e-waste recycling sites in addition to toxic metallic compounds. E-waste recycling sites are hence prime examples of organic/metallic chemical mixture exposure situations in both occupational and environmental contexts and may cause important health outcomes. The issue of e-waste chemical mixtures and interactions between chemicals in relation to public health risk assessments is discussed in the following sections.

4. CHEMICAL MIXTURES EXPOSURES IN E-WASTE RECYCLING

As noted above, human exposure to chemicals in e-waste materials will occur as chemical mixtures. These exposures will occur from combinations of a number of metallic compounds and common organic chemicals/plastics released during open-pit burning and dumping not only into arable soils used for growing crops but also into rivers, lakes, and shallow water embayments used for harvesting edible fish and shellfish. Human exposures to these chemicals may hence occur via a number of routes over the lifetime of individuals living in e-waste recycling areas. Since a number of these chemicals are capable of crossing the placenta, the exposures may occur prior to conception, through embryogenesis and fetal development leading to health effects that manifest themselves later in life (Grant et al., 2013) due to altered cellular imprinting (Murphy & Jirtle, 2003; Wilkinson, Davies, & Isles, 2007). In general, interactions among chemicals may occur as additive, antagonistic, or synergistic in nature (Fowler et al., 2005, 2008; Mahaffey et al., 1981; Fig. 2.1) and are not always easily predicable. Further complicating matters is the increasing use of nanomaterials in the production of electronic devices (Cui & Lieber, 2001; Miao, Miyauchi, Simmons, Dordick, & Linhardt, 2010). These materials may greatly alter the absorption, distribution, and elimination of their constituent chemical components, thus further complicating risk assessment predictions.

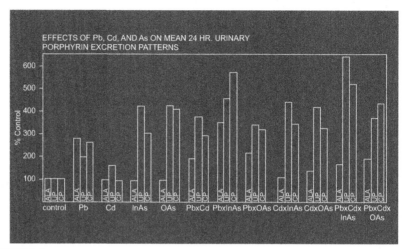

FIGURE 2.1

Urinary porphyrin excretion patterns from rats exposed to inorganic arsenic (As), lead (Pb), cadmium, or organic arsenic compounds (OAs) as arsanilic acid on an individual or mixture basis. *Interactions Fowler, B. A., & Mahaffey, K. R. (1978). Interaction between lead, cadmium and arsenic in relation to porphyrin excretion patterns.* Environmental Health Perspectives, 25, *87–90.*

5. RISK ASSESSMENT APPROACHES FOR E-WASTE

Given the growing, unique, and evolving nature of the global e-waste problem, it would appear that traditional approaches to chemicals risk assessment will not be sufficient to protect humans or the environment from the untoward effects of chemicals released from e-waste materials during recycling activities. This is particularly true for recycling processing conducted in developing countries with few, if any, environmental or occupational safeguards or child labor laws. Clearly, the challenges of e-waste recycling require the application of newer risk assessment methods capable of evaluating the effects of novel inorganic and organic e-waste chemicals on an individual and mixture basis in populations at special risk in developing countries. These populations may be defined on the basis of age (Fowler, 2013a), gender (Fowler et al., 2005), genetic inheritance (Scinicariello et al., 2007; Fig. 2.2), and nutritional status (Heindel & vom Saal, 2009), resulting in the need for risk assessment approaches capable of integrating these disparate factors to provide credible mode of action (MOA)–based guidance that will hopefully ultimately lead to the development of personalized risk assessment evaluations for sensitive subpopulations at special risk for adverse outcomes.

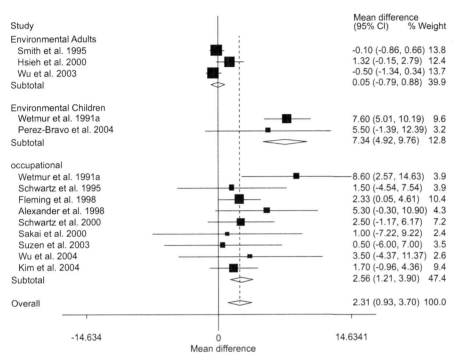

FIGURE 2.2

Metaanalysis of relationships between blood lead values and ALAD1/2 genotypes. *From Scinicariello, F., Murray, H. E., Moffett, D. B., Abadin, H., Sexton, M. J., & Fowler, B. A. (2007). Lead and δ-aminolevulinic acid dehydratase polymorphism: Where does it lead? A meta-analysis.* Environmental Health Perspectives, 115, 35–41.

Fortunately, there are a number of modern toxicological tools such as molecular biomarkers and computational methodologies that have evolved over the past 20 years, which may be able to address these complex issues if properly applied.

5.1 Molecular Biomarkers

A prior book in this series (Fowler, 2016) has provided an overview of some ways in which molecular biomarkers could be applied to provide more precise risk assessments for chemical exposures arising from e-waste recycling. Biomarker test batteries are becomingly increasingly automated and less expensive so that they may become available for general screening of chemical toxicity in developing countries where a large proportion of e-waste recycling is currently centered (Sthiannopkao & Wong, 2013). There are an increasing array of molecular biomarkers, which include genomic, proteomic, and metabolomics biomarkers, which are discussed in more detail elsewhere (Fowler, 2016).

These biomarkers either alone or in combination can provide useful insights into mechanisms of toxic chemical action and hence help to support MOA risk assessment practice (Cote et al., 2016).

5.2 Computational Toxicology Methodologies

A second essential component for providing cost-effective personalized risk assessments for e-waste chemical exposures is the application of computational toxicology methods capable of integrating large quantities of diverse data to generate an overall risk assessment picture of likely adverse health outcomes in exposed populations at special risk. Such credible information is necessary to formulate wise regulatory decisions. Fortunately, the field of computational toxicology has also evolved in recent years. A relatively recent summary of this field and its potential applications has been published (Fowler, 2013b) as a component of the present three-part series on the application of modern tools of toxicology to risk assessment practice. An example of the value of computational toxicology as a tool in helping to drive MOA risk assessment is via digital image analysis of 2D gels from male or female hamsters exposed to gallium arsenide or indium arsenide particles (Fowler et al., 2005). The images were converted into tabular formats and clearly show marked gender-specific differences in response patterns with regard to gallium arsenide or indium arsenide particles at equal dose levels (see Tables 2.1 and 2.2). These data highlight the value of computational techniques in supporting risk assessments for chemical mixtures on a gender-specific basis. This book is the third volume in this series and represents a real-world case study of how the information in the first two volumes could be applied for addressing the complex and growing problem of e-waste with particular emphasis on the public health aspects.

6. PUBLIC HEALTH IMPLICATIONS AND DIRECTIONS FORWARD

It is clear from the above summary that unregulated releases of toxic metals from e-waste recycling are occurring and that human exposures from air, food, and water are occurring. This situation is further complicated by exposures of children in developing countries without child labor laws and the expanded application of nanomaterials in electronic devices, which can only increase environmental dispersions and exposures of humans and other biota.

6.1 The Current Situation

Based on the brief review of the various aspects of the global e-waste problems stated above, it is clear that there is a growing public health problem with human exposures to chemicals, such as toxic metals, derived from unregulated recycling

Table 2.1 Polypeptides That Exhibit Modulation 30 Days Following Exposure to InAs and GaAs in Hamster Kidney Proximal Tubule Cells

MW Range	Common Spot Number[a]	InAs[b]	GaAs[b]
100–90	1	1.0	2.1[c]
	18	1.0	0.7
89–70	2	–	–
	12	–	1.1
69–50	3	0.90	4.0[c]
	11	–	1.2
	22	–	–
	23	–	–
	25	–	–
	32	–	–
49–40	10	0.85	1.8
	16	1.2	1.6
	26	–	–
	28	–	–
39–30	4	0.79	2.0[c]
	5	1.8	0.5[c]
	6	–	0.1[c]
	7	–	0.3[c]
	8	1.5	3.6[c]
	13	0.9	0.1[c]
	14	0.72	0.6
	15	1.2	1.7
	17	0.74	2.4[c]
	19	–	–
	20	–	–
29–20	30	–	–
	31	–	–

MW, molecular weight.
[a]Spot number denoted on gel.
[b]Spot intensity expressed as the ratio of treatment/control.
[c]Denotes polypeptides differing by twofold or greater (increasing or decreasing).
From Fowler, B. A., Conner, E. A., & Yamauchi, H. (2008). Proteomic and metabolomic bio-markers for III-V semiconductors: And prospects for application to nano-materials. Toxicology and Applied Pharmacology, 233(1), 110–115. http://dx.doi.org/10.1016/j.taap.2008.01.014.

of electronic devices, which is centered in a number of developing countries. These countries generally do not have the resources or political will to put in place needed regulatory guidelines for protecting the health of citizens dealing with e-waste materials. A key element in improving this situation is the availability of sound scientific information to inform regulatory decision-making.

Table 2.2 Polypeptides That Exhibit Modulation 30 Days Following InAs or GaAs in Female Hamster Kidney Proximal Tubule Cells

MW Range	Common Spot Number[a]	InAs[b]	GaAs[b]
100–90	1	0.92	0.96
	18	0.56	0.17[c]
	29	0.71	0.37[c]
89–70	2	6.60[c]	5.50[c]
	12	0.47[c]	1.20
69–50	3	0.68	1.10
	11	0.51	0.96
	22	0.32[c]	0.78
	23	0.31[c]	0.83
	24	0.26[c]	0.64
	25	0.26[c]	1.26
	32	0.90	0.96
49–40	10	0.43[c]	0.90
	16	2.50[c]	0.57
	26	0.58	0.45[c]
	27	–	1.60
	28	–	10.0[c]
	33	–	1.40
	34	0.43[c]	0.57
39–30	4	–	–
	5	2.00[c]	1.20
	6	0.14[c]	0.70
	7	0.32[c]	0.61
	8	0.39[c]	0.50
	13	0.47[c]	1.20
	14	0.37[c]	1.00
	15	0.76	1.25
	17	1.00	0.96
	19	0.20[c]	1.1
	20	–	0.50
≤29	30	0.84	0.76
	31	0.54	0.68

MW, molecular weight.
Thirty-one polypeptides were affected by both InAs or GaAs. The synthesis of 16 polypeptides was altered by InAs by twofold or greater with the synthesis of 3 increasing, 13 decreasing, and 4 absent. After GaAs treatment, 50% of the polypeptides were synthesized at or near control levels. Five polypeptides were changed by twofold or greater, the synthesis of two increased, three decreased, and one was absent.
[a]Spot number denoted on gel.
[b]Spot intensity expressed as the ratio of treatment/control.
[c]Denotes polypeptides differing by twofold or greater (increasing or decreasing).
From Fowler, B. A., Conner, E. A., & Yamauchi, H. (2008). Proteomic and metabolomic biomarkers for III-V semiconductors: And prospects for application to nano-materials. Toxicology and Applied Pharmacology, 233(1), 110–115. http://dx.doi.org/10.1016/j.taap.2008.01.014.

6.2 Directions Forward

To make progress on this complex problem area and protect the health of the environment and the public, a number of elements need to be organized in concert.

1. Training and resources need to be marshaled to provide recycling environments that are more efficient and protect the workers and minimize releases of toxic chemicals from electronic devices during the recycling process.
2. Occupational and environmental laws need to be strengthened to encourage safe recycling practices in developing countries.
3. Financial incentives in the form of tax breaks or subsidies need to be provided to companies engaged in the manufacture or recycling of electronic devices to encourage safe recycling and/or refurbishing practices.
4. The utilization of modern approaches to toxicology and risk assessment should be included for evaluation of chemical safety during any phase of e-waste handling to assure that public health is being protected for the most sensitive segments of the population.
5. To be effective and have an impact on public health in developing countries engaged in e-waste recycling, the proposed newer methods must be affordable and cost-effective in developing countries to have an impact on public health issues related to e-waste chemicals. Fortunately, the costs of these evolving tests are declining every year due to incorporation of computer-managed analytical and data management systems. It is reasonable to expect that such evaluations will be practical in even remote areas via incorporation of satellite data telemetry systems to communicate biomarker-based risk assessment data to risk assessors located in more centralized urban locations.

7. SUMMARY AND CONCLUSIONS

From the above discussion, it is clear that metallic elements play essential roles in modern electronic devices and are hence important components of e-waste streams. Some of these elements such as gold are regarded as precious metals and there is considerable effort exerted to recover them. Other metals such as indium and gallium are not only valuable components of semiconductors but also show considerable toxic potential. Elements such as arsenic, cadmium, lead, and mercury are well-known toxicants but financially less valuable, so less effort is exerted to recover them and hence they frequently find their way into air, food, and water where they may produce toxicity on an individual or mixture basis. From the risk assessment perspective, the situation for metallic elements as a class in e-waste is complicated in terms of conducting an accurate evaluation.

This situation is further complicated by concomitant exposures to organic chemicals such as those released by burning insulation from copper wiring and combustion of plastics and flame retardant chemicals. The issue of chemical interactions between metals and organic chemicals in mixture situations is difficult from a risk assessment perspective since these agents may affect different pathways leading to cell injury/cell death or cancer outcomes. In addition, there are biological factors that should be considered in conducting more accurate risk assessments for populations exposed to chemicals released from e-waste recycling activities in developing countries. These include age—the developing fetus exposed to chemicals capable of crossing the placenta, children working in e-waste recycling activities—gender, and intrinsic differences in susceptibility to metal toxicity (Fowler et al., 2005, 2008), and nutritional status. Persons with poor nutritional status are generally less resistant to the effects of toxic chemicals than persons with good nutritional status (Heindel & vom Saal, 2009). Other biological factors such as the presence of infectious diseases (Ortiz et al., 2002) may also play a role in health outcomes since a number of e-waste chemicals produce immune-suppressive effects (Luster et al., 1992). Finally, there is the impact of low socioeconomic status (SES) itself on susceptibility to chemical-induced diseases (Friedman & Lawrence, 2002). Low SES is a major driver for persons in developing countries to engage in the unregulated recycling of e-waste. It is well known that low SES itself is major determinant of decreased longevity (Bassuk, Berkman, & Amick, 2002) for a complex set of reasons, and the impact of this factor coupled with exposure to e-waste chemicals should be considered in future risk assessment paradigm. A diagrammatic overview representation of such an integrative risk assessment approach is presented in Figs. 2.3 and 2.4.

COLLECTION OF DISCARDED ELECTRONIC DEVICES (E-WASTE)

|

CONTAINER SHIP TRANSPORT OF E- WASTE TO DEVELOPING COUNTRIES

|

TRANSFER OF E-WASTE FROM SHIPPING PORT VIA TRUCK TO UN-REGULATED RE-CYCLING SITES

|

DISMANTLING OF E-WASTE DEVICES AND BREAKING OF HARD DRIVES AND CIRCUIT BOARDS BY HAND

TOOLS

| - HUMAN EXPOSURES TO PARTICLES AND DUST

MELTING OF METAL AND PLASTIC COMPONENTS IN OPEN AIR CONDITIONS TO RECOVER PRECIOUS

METALS WITH VOLATILE RELEASES OF TOXIC METALS AND PLASCTIC COMPONENT CHEMICALS

| - HUMAN EXPOSURES TO VOLATILIZED TOXIC METALLICS

AND CHEMICALS FROM PLASTICS

STRIPPING OF INSULATION FROM COPPER WIRNG BY MELTING IN OPEN AIR CONDITIONS WITH RELEASE

OF INSULATION CHEMICALS

| – HUMAN EXPOSURES TO VOLATILIZED INSULATION

CHEMICALS

|

HUMAN EXPOSURES FOLLOWING DISPOSAL OF UNSALVAGEABLE E-WASTE MATERIALS IN LANDFILLS OR

WATER BODIES – CHRONIC RELEASES OF TOXICS CHEMICALS WITH UPTAKE INTO CROPS AND EDIBLE

FISHERY PRODUCTS AND HUMAN EXPOSURE FOLLOWING INGESTION

KNOWN POPULATION RISK ASSESSMENT FACTORS

AGE |

GENDER

GENETIC SUSCEPTIBILITY   >

NUTRITIONAL STATUS

SES |

COLLECT EPIDEMIOLOGICAL DATA ON POPULATION OF INTEREST

FIGURE 2.3

A diagrammatic representation of a possible integrative risk assessment paradigm for evaluating potential health effects from e-waste chemicals in susceptible populations in developing countries.

COLLECT ACTUAL HUMAN EXPOSURE DATA VIA BIOMONITORING STUDIES OF ACCESSIBLE MATRICES

(BLOOD, HAIR, URINE, TISSUE SWABS)

|

DATA MINING OF PUBLISHED AND UNPUBLISHED (GRAY) LITERATURE ON KNOWN

CHEMICALS IN SITE MIXTURES

|

"BIN" KNOWN CHEMICALS BY RANKING THEM FROM MOST TOXIC TO LESS TOXIC FILL DATA GAPS FOR

CHEMICALS WITH UNKNOWN TOXIC POTENTIAL USING QSAR/PBPK/MOLECULAR DOCKING

TECHNIQUES

|

CONDUCT MIXTURE RISK ASSESSMENT CALCULATIONS USING

1)   MOST TOXIC KNOWN CHEMICAL APPROACH

OR

2)   INTEGRATE KNOWN OR COMPUTATIONAL MODELING - PREDICTED TOXICITIES OF CHEMICALS

IN THE MIXTURE USING AN ADDITIVITY FORMAT* TO DERIVE AN OVERALL RISK ASSESSMENT

SCORE FOR THE MIXTURE

*NOTE-THE ASSUMPTION OF ADDITIVITY IS RECOMMENDED FOR MIXTURE ASSESSMENTS BASED

ON PUBLISHED DATA INDICATING THAT IT IS THE MOST COMMON TYPE OF INTERACTION BETWEEN

CHEMICALS AND IT IS A CONSERVATIVE APPROACH WHICH ALSO ADDRESSES THE ISSUE OF

ANTAGONISM BETWEEN CHEMICALS. THE ISSUE OF POSSIBLE SYNERGISTIC INTERACTIONS

BETWEEN CHEMICALS IS DIFFICULT TO PREDICT WITHOUT ACTUAL EXPERIMENTAL DATA TO GUIDE

THE CALCULATIONS BUT COULD BE COVERD BY STATISTICAL PROBALISTIC RISK ASSESSMENT

APPROACHES

FIGURE 2.4

Diagammatic approach for conducting risk assessments on e-waste chemical mixture exposures among human populations in developing countries.

References

Aoki, Y., Lipsky, M. M., & Fowler, B. A. (1990). Alteration of protein synthesis in primary cultures of rat kidney epithelial cells by exposure to gallium indium and arsenite. *Toxicology and Applied Pharmacology, 106*, 462–468.

Bassuk, S. S., Berkman, L. F., & Amick, B. C. (2002). Socioeconomic status and mortality among the elderly: Findings from four US communities. *American Journal of Epidemiology, 155*(6), 520–533.

Bustamente, J., Dock, L., Vahter, M., Fowler, B., & Orrenius, S. (1997). The semiconductor elements arsenic and indium induce apoptosis in rat thymocytes. *Toxicology, 118*, 129–136.

Caballero-Guzman, A., Sun, T., & Nowack, B. (2015). Flows of engineered nanomaterials through the recycling process in Switzerland. *Waste Management, 36*, 33–43.

Conner, E. A., Yamauchi, H., & Fowler, B. A. (1995). Alterations in the heme biosynthetic pathway from III-V semiconductor metal indium arsenide (InAs). *Chemico-Biological Interactions, 96*, 273–285.

Cote, I., Andersen, M. E., Ankley, G. T., Barone, S., Birnbaum, L. S., Boekelheide, K., & DeWoskin, R. S. (2016). The next generation of risk assessment multiyear study–highlights of findings, applications to risk assessment and future directions. *Environmental Health Perspectives.* http://dx.doi.org/10.1289/ehp233.

Cui, Y., & Lieber, C. M. (2001). Functional nanoscale electronic devices assembled using silicon nanowire building blocks. *Science, 291*(5505), 851–853.

Cui, J., & Zhang, L. (2008). Metallurgical recovery of metals from electronic waste: A review. *Journal of Hazardous Materials, 158*(2), 228–256.

Fowler, B. A. (2013a). Cadmium and aging. In B. Weiss (Ed.), *Aging and vulnerabilities to environmental chemicals. Royal society of chemistry* (pp. 376–387). UK: Cambridge.

Fowler, B. A. (Ed.). (2013b). *Computational toxicology: Applications for risk assessment.* pp. 258. Amsterdam: Elsevier Publishers.

Fowler, B. A. (2016). *Molecular biological markers for toxicology and risk assessment.* Amsterdam: Elsevier Publishers, pp. 153.

Fowler, B. A., Conner, E. A., & Yamauchi, H. (2005). Metabolomic and proteomic biomarkers for III-V semiconductors: Chemical-specific porphyrinurias and proteinurias. *Toxicology and Applied Pharmacology, 206*(2), 121–130. http://dx.doi.org/10.1016/j.taap.2005.01.020.

Fowler, B. A., Conner, E. A., & Yamauchi, H. (2008). Proteomic and metabolomic biomarkers for III-V semiconductors: And prospects for application to nano-materials. *Toxicology and Applied Pharmacology, 233*(1), 110–115. http://dx.doi.org/10.1016/j.taap.2008.01.014.

Fowler, B. A., & Mahaffey, K. R. (1978). Interaction between lead, cadmium and arsenic in relation to porphyrin excretion patterns. *Environmental Health Perspectives, 25*, 87–90.

Friedman, E. M., & Lawrence, D. A. (2002). Environmental stress mediates changes in neuroimmunological interactions. *Toxicological Sciences, 67*(1), 4–10.

Goering, P. L., Maronpot, R. R., & Fowler, B. A. (1988). Effect of intratracheal administration of gallium arsenide administration on δ-aminolevulinic acid dehydratase in rats: Relationship to urinary excretion of aminolevulinic acid. *Toxicology and Applied Pharmacology, 92*, 179–193.

Grant, K., Goldizen, F. C., Sly, P. D., Brune, M.-N., Neira, M., van den Berg, M., & Norman, R. E. (2013). Health consequences of exposure to e-waste: A systematic review. *The Lancet Global Health, 1*(6), e350–e361.

Heeres, E. C., Bakkers, E. P., Roest, A. L., Kaiser, M., Oosterkamp, T. H., & de Jonge, N. (2007). Electron emission from individual indium arsenide semiconductor nanowires. *Nano Letters, 7*(2), 536–540.

Heindel, J. J. (2003). Endocrine disruptors and the obesity epidemic. *Toxicological Sciences, 76*(2), 247–249.

Heindel, J. J., & vom Saal, F. S. (2009). Role of nutrition and environmental endocrine disrupting chemicals during the perinatal period on the aetiology of obesity. *Molecular and Cellular Endocrinology, 304*(1), 90–96.

Huang, D.-Y., Zhao, H.-Q., Liu, C.-P., & Sun, C.-X. (2014). Characteristics, sources, and transport of tetrabromobisphenol A and bisphenol A in soils from a typical e-waste recycling area in South China. *Environmental Science and Pollution Research, 21*(9), 5818–5826.

Jirtle, R. L., & Skinner, M. K. (2007). Environmental epigenomics and disease susceptibility. *Nature Reviews Genetics, 8*(4), 253–262.

Kiran, N., Ekinci, E., & Snape, C. (2000). Recyling of plastic wastes via pyrolysis. *Resources, Conservation and Recycling, 29*(4), 273–283.

Klaine, S. J., Alvarez, P. J., Batley, G. E., Fernandes, T. F., Handy, R. D., Lyon, D. Y., … Lead, J. R. (2008). Nanomaterials in the environment: Behavior, fate, bioavailability, and effects. *Environmental Toxicology and Chemistry, 27*(9), 1825–1851.

Landi, B., Castro, S., Ruf, H., Evans, C., Bailey, S., & Raffaelle, R. (2005). CdSe quantum dot-single wall carbon nanotube complexes for polymeric solar cells. *Solar Energy Materials and Solar Cells, 87*(1), 733–746.

Leung, A. O., Duzgoren-Aydin, N. S., Cheung, K., & Wong, M. H. (2008). Heavy metals concentrations of surface dust from e-waste recycling and its human health implications in southeast China. *Environmental Science and Technology, 42*(7), 2674–2680.

Lowry, G. V., Gregory, K. B., Apte, S. C., & Lead, J. R. (2012). Transformations of nanomaterials in the environment. *Environmental Science and Technology, 46*(13), 6893–6899.

Luster, M. I., Portier, C., Paît, D. G., White, K. L., Gennings, C., Munson, A. E., & Rosenthal, G. J. (1992). Risk assessment in immunotoxicology I. Sensitivity and predictability of immune tests. *Toxicological Sciences, 18*(2), 200–210.

Madden, E. F., & Fowler, B. A. (2000). Mechanisms of nephrotoxicity from metal combinations: A review. *Drug and Chemical Toxicology, 23*, 1–12.

Mahaffey, K. R., Capar, S. G., Gladen, B. C., & Fowler, B. A. (1981). Concurrent exposure to lead, cadmium, and arsenic. Effects on toxicity and tissue metal concentrations in the rat. *The Journal of Laboratory and Clinical Medicine, 98*(4), 463–481.

Matsukami, H., Tue, N. M., Suzuki, G., Someya, M., Viet, P. H., Takahashi, S., … Takigami, H. (2015). Flame retardant emission from e-waste recycling operation in Northern Vietnam: Environmental occurrence of emerging organophosphorus esters used as alternatives for PBDEs. *Science of the Total Environment, 514*, 492–499.

Miao, J., Miyauchi, M., Simmons, T. J., Dordick, J. S., & Linhardt, R. J. (2010). Electrospinning of nanomaterials and applications in electronic components and devices. *Journal of Nanoscience and Nanotechnology, 10*(9), 5507–5519.

Murphy, S. K., & Jirtle, R. L. (2003). Imprinting evolution and the price of silence. *Bioessays, 25*(6), 577–588.

Mushonga, P., Onani, M. O., Madiehe, A. M., & Meyer, M. (2012). Indium phosphide-based semiconductor nanocrystals and their applications. *Journal of Nanomaterials, 2012*, 12.

Ni, H. G., Zeng, H., Tao, S., & Zeng, E. Y. (2010). Environmental and human exposure to persistent halogenated compounds derived from e-waste in China. *Environmental Toxicology and Chemistry, 29*(6), 1237–1247.

Ortiz, D., Calderón, J., Batres, L., Carrizales, L., Mejía, J., Martínez, L., … Díaz-Barriga, F. (2002). Overview of human health and chemical mixtures: Problems facing developing countries. *Environmental Health Perspectives, 110*(Suppl. 6), 901.

Perera, F., & Herbstman, J. (2011). Prenatal environmental exposures, epigenetics, and disease. *Reproductive Toxicology, 31*(3), 363–373.

Robinson, B. H. (2009). E-waste: An assessment of global production and environmental impacts. *Science of the Total Environment, 408*(2), 183–191.

Scinicariello, F., Murray, H. E., Moffett, D. B., Abadin, H., Sexton, M. J., & Fowler, B. A. (2007). Lead and δ-aminolevulinic acid dehydratase polymorphism: Where does it lead? A meta-analysis. *Environmental Health Perspectives, 115*, 35–41.

Sepúlveda, A., Schluep, M., Renaud, F. G., Streicher, M., Kuehr, R., Hagelüken, C., & Gerecke, A. C. (2010). A review of the environmental fate and effects of hazardous substances released from electrical and electronic equipments during recycling: Examples from China and India. *Environmental Impact Assessment Review, 30*(1), 28–41.

Shen, C., Chen, Y., Huang, S., Wang, Z., Yu, C., Qiao, M., … Lin, Q. (2009). Dioxin-like compounds in agricultural soils near e-waste recycling sites from Taizhou area, China: Chemical and bioanalytical characterization. *Environment International, 35*(1), 50–55.

Skinner, M. K., Manikkam, M., & Guerrero-Bosagna, C. (2011). Epigenetic transgenerational actions of endocrine disruptors. *Reproductive Toxicology, 31*(3), 337–343.

Sthiannopkao, S., & Wong, M. H. (2013). Handling e-waste in developed and developing countries: Initiatives, practices, and consequences. *Science of the Total Environment, 463*, 1147–1153.

Tang, X., Hashmi, M. Z., Long, D., Chen, L., Khan, M. I., & Shen, C. (2014). Influence of heavy metals and PCBs pollution on the enzyme activity and microbial community of paddy soils around an e-waste recycling workshop. *International Journal of Environmental Research and Public Health, 11*(3), 3118–3131. http://dx.doi.org/10.3390/ijerph110303118.

Vejerano, E. P., Leon, E. C., Holder, A. L., & Marr, L. C. (2014). Characterization of particle emissions and fate of nanomaterials during incineration. *Environmental Science: Nano, 1*(2), 133–143.

Walser, T., Limbach, L. K., Brogioli, R., Erismann, E., Flamigni, L., Hattendorf, B., … Stark, W. J. (2012). Persistence of engineered nanoparticles in a municipal solid-waste incineration plant. *Nature Nanotechnology, 7*(8), 520–524.

Whittaker, M., Wang, G., Chen, X.-Q., Lipsky, M., Smith, D., Gwiazda, R., & Fowler, B. A. (2011). Effects of trace element mixtures and the production of oxidative stress precursors: 30-day, 90-day, and 180-day drinking water studies in rats. *Toxicology and Applied Pharmacology, 254*(2), 154–166.

Wilkinson, L. S., Davies, W., & Isles, A. R. (2007). Genomic imprinting effects on brain development and function. *Nature Reviews Neuroscience, 8*(11), 832–843.

Wong, C. S., Duzgoren-Aydin, N. S., Aydin, A., & Wong, M. H. (2007). Evidence of excessive releases of metals from primitive e-waste processing in Guiyu, China. *Environmental Pollution, 148*(1), 62–72.

Wong, M., Wu, S., Deng, W., Yu, X., Luo, Q., Leung, A., … Wong, A. (2007). Export of toxic chemicals – A review of the case of uncontrolled electronic-waste recycling. *Environmental Pollution, 149*(2), 131–140.

Zhao, G., Zhou, H., Wang, D., Zha, J., Xu, Y., Rao, K., … Wang, Z. (2009). PBBs, PBDEs, and PCBs in foods collected from e-waste disassembly sites and daily intake by local residents. *Science of the Total Environment, 407*(8), 2565–2575.

Zheng, J., Chen, K.-H., Yan, X., Chen, S.-J., Hu, G.-C., Peng, X.-W., … Yang, Z.-Y. (2013). Heavy metals in food, house dust, and water from an e-waste recycling area in South China and the potential risk to human health. *Ecotoxicology and Environmental Safety, 96*, 205–212.

Zhuo, C., & Levendis, Y. A. (2014). Upcycling waste plastics into carbon nanomaterials: A review. *Journal of Applied Polymer Science, 131*(4).

CHAPTER 3

Toxicology of E-Waste Chemicals—Mechanisms of Action

INTRODUCTION

1. TOXIC METALS/METALLOIDS

Electron waste contains a number of toxic chemicals, which can be divided into metallic elements and organic chemicals. A number of these agents have been extensively studied, and there is solid toxicological database for understanding their potential hazard. Others have not been so well studied, and there is the issue of mixture exposures. Each general category of e-waste chemicals has its own special set of concerns with regard to knowledge base and hence being amenable to mode of action (MOA) risk assessment approaches. This first section of this chapter will focus on toxic metals and metalloids on an individual or compound basis since both may be present in e-waste materials. It should be noted here that some of these elements such as gold and indium are valuable, leading to extensive efforts for their recovery. Other toxic elements such as lead, cadmium, chromium, and arsenic are less valuable and may be released in both occupational exposures and into the general environment. Plastics and organic chemicals in electronic devices, such as those incorporated into electronics as fire retardants, are less valuable and frequently released during the recycling process. This is a particularly important issue in developing countries with limited resources for occupational or environmental protection. Finally, the chapter will attempt to briefly summarize the sources of exposures and mechanisms of toxicity for some of the known major toxic inorganic and organic e-waste chemicals and highlight populations at special risk for toxic outcomes.

1.1 Lead

Lead, with atomic number 82, whose toxic properties have been known for centuries (NAS/NRC, 1993) has been used as shielding in cathode ray tubes (Herat, 2008) and solders for circuit boards (Li, Lu, Guo, Xu, & Zhou, 2007; Suganuma, 2001). Humans could be exposed to lead dust from the destruction of these common e-waste components via lead-containing dust and lead fumes from breakage and incineration activities (Wong et al., 2007). This is a particular issue for the fetus, children, and women of childbearing age (Meyer, Brown, & Falk, 2008); however, all age groups and a number of organ systems including the nervous system, kidneys, blood-forming organs, reproductive systems (Needleman, 2004), and the cardiovascular system(Navas-Acien, Guallar, Silbergeld, & Rothenberg, 2007) may be affected. Nutritional status (Hsu & Guo, 2002) and genetic inheritance (Onalaja & Claudio, 2000; Scinicariello, Yesupriya, Chang, & Fowler, 2010) may also play important roles in defining sensitive subpopulations at special risk for toxicity.

The mechanisms of lead toxicity in target organs seem to be complex and markedly influenced by the handling of this element by the skeleton (NAS/NRC, 1993) and intracellular lead-binding proteins (Fowler, 1998) and intranuclear inclusion bodies at elevated exposure levels (Oskarsson & Fowler, 1985a). The biologically available intracellular fraction of lead may interact with a number of organelle systems including the nucleus, mitochondria, and cytosolic fractions (Oskarsson & Fowler, 1985a,b) with resultant perturbations of a number of essential cellular functions (Oskarsson & Fowler, 1985b; Shelton, Todd, & Egle, 1986).

1.2 Cadmium

Cadmium is another toxic element used in semiconductor industry in solders (Dolzhnikov et al., 2015) and more recently in II–VI semiconductor materials (Adachi, 2009). This element has been classified as a Class I carcinogen by IARC (2006) and (Waalkes, 2000) is well known to produce toxicity in kidneys (Nordberg, Fowler, & Nordberg, 2015), the skeleton (Takebayashi, Jimi, Segawa, & Kiyoshi, 2000), and reproductive organs (Akinloye, Arowojolu, Shittu, & Anetor, 2006; Thompson & Bannigan, 2008) via a number of direct and indirect mechanisms (Fowler, 2009; Klaassen, Liu, & Diwan, 2009; Prozialeck, 2000; Waisberg, Joseph, Hale, & Beyersmann, 2003). All age groups may be susceptible but females (Nishijo, Satarug, Honda, Tsuritani, & Aoshima, 2004; Ruiz, Mumtaz, Osterloh, Fisher, & Fowler, 2010; Tellez-Plaza, Navas-Acien, Crainiceanu, Sharrett, & Guallar, 2010; Vahter, Åkesson, Lidén, Ceccatelli, & Berglund, 2007), multiparous postmenopausal women (Bhattacharyya, 1991; Kazantzis, 2004), and the elderly (Fowler, 2013a) seem to be at special risk for adverse outcomes. The cysteine-rich protein metallothionein (MT) plays a major role in the handling of cadmium in tissues. This protein exists as a number of isoforms, which vary between various tissues (Cherian, Jayasurya, & Bay, 2003; Thirumoorthy, Sunder, Kumar, Ganesh, & Chatterjee, 2011). In general, MT appears to modulate both the transport of Cd in the circulation and the intracellular bioavailability of Cd within cells (Squibb, Pritchard, & Fowler, 1984). Once the intracellular binding capacity of MT is exceeded and Cd^{2+} ions are available to interact with sensitive sites, more overt manifestations of cell death are initiated (Squibb et al., 1984).

Cadmium is also a potent initiator of oxidative stress via generation of reactive oxygen species (ROS) (Szuster-Ciesielska et al., 2000; Wang, Fang, Leonard, & Rao, 2004). These ROS are capable of altering normal signaling pathways and produce a number of effects on hormone systems such as those involved in reproduction (Chedrese, Piasek, & Henson, 2006; Safe, 2003; Takiguchi & Yoshihara, 2005). The mechanisms involved in these effects are complex since cadmium itself should not catalyze Fenton chemistry, and hence interference with cellular oxidation/reduction systems and/or depletion of intracellular antioxidant systems are more likely the causes.

1.3 Arsenic

The element arsenic, which is found in a wide variety of electronic devices, is of particular concern, and it is not as valuable as some of the other elements discussed below, so efforts to recover it during recycling are less rigorous. This element, which exists in three main oxidation states (+/−3, +5), may be volatilized by high temperatures creating both potential occupational and environmental hazards (Fawcett & Jamieson, 2011; Henke, 2009). These main oxidation states vary in their acute toxic potential (Fowler, 2013b). In addition, inorganic arsenicals may be methylated to form a variety of methylated species (monomethyl arsenic acids, dimethyl arsenic acids, and trimethyl arsines), which also vary in their relative toxicity (Fowler, 2015; Styblo et al., 2000), and it is possible that intracellular toxicity may be due in part to metabolic interconversions among these methylated species (Aposhian, Zakharyan, Avram, Sampayo-Reyes, & Wollenberg, 2004; Thomas et al., 2007). The mechanisms by which arsenicals produce toxicity seem to be largely centered around effects on inhibition of cellular respiration (Samikkannu et al., 2003) with resultant generation of ROS (Samikkannu et al., 2003). An excess of ROS can in turn produce oxidative stress (Flora, 2011), proteotoxicity (Bolt, Zhao, Pacheco, & Klimecki, 2012; Stanhill et al., 2006), and initiation of apoptotic and necrosis cell death pathways (Bustamante, Nutt, Orrenius, & Gogvadze, 2005) and initiation of arsenic-induced carcinogenesis (Shi, Hudson, & Liu, 2004; Shi, Shi, & Liu, 2004). The combined effects of arsenic with other toxic elements such as gallium and indium in III–V semiconductors such as gallium arsenide and indium arsenide are discussed below.

1.4 Mercury

Mercury is a well-known toxic element that can exist in the 0, +1, or +2 oxidation states and as a number of alkylated forms such as methylmercury, dimethylmercury, and ethylmercury and a number of ring structured forms (Clarkson & Magos, 2006), which vary in their uptake and distribution (Clarkson, Vyas, & Ballatori, 2007). Mercury is used in electronic devices such as flat panel televisions and LCDs (Lim & Schoenung, 2010) and switches (Babu, Parande, &

Basha, 2007) and may be released as Hg^0 vapor during the recycling process. This volatile form may contribute to both occupational exposures during recycling and environmental exposures following microbial methylation reactions (Parks et al., 2013; Ullrich, Tanton, & Abdrashitova, 2001). Methylmercury is the chemical form of greatest environmental concern because of its ability to accumulate in large predator fish species (García-Hernández et al., 2007; Hightower & Moore, 2003; Oken et al., 2003), which can hence lead to human exposures from this food source. The in vivo metabolism of organomercurials is complex and may involve both dealkylation reactions to form inorganic mercury (Suda, Suda, & Hirayama, 1993), which is a potent inducer of MT (Tandon, Singh, Prasad, & Mathur, 2001; Yasutake, Nakano, & Hirayama, 1998), and alkylation reactions mediated by bacterial flora in the microbiome (Betts, 2011; Podar et al., 2015), leading to the formation of methylmercury species. A major point to be noted here is that all of these chemical forms of mercury are toxic to biological systems to some degree. The mechanisms of mercurial toxicity are also complex since these agents may affect a number of essential subcellular systems including the mitochondria (Fowler & Woods, 1977; Lund, Miller, & Woods, 1991), protein synthetic machinery (Nakada, Nomoto, & Imura, 1980; Syversen, 1981; Verity, Brown, Cheung, & Czer, 1977), and cell death pathways (Shenker, Guo, & Shapiro, 1998). The alkylated forms of mercury are a particular problem because of their lipophilic nature and ability to cross cellular and intracellular membranes and penetrate virtually every compartment of the cell (Norseth & Brendeford, 1971).

1.5 Gallium

Gallium is a commonly used element in the production of electronic devices such as computer chips, cellular telephones, and light-emitting diodes (LEDs) (Fowler & Sexton, 2015; Moskalyk, 2003; Rajan & Jena, 2013), and it is recovered as a by-product of aluminum and zinc smelting (Moskalyk, 2003). This element exists in the +3 oxidation state, and metabolism of gallium in vivo seems to be similar to that of iron since administration of gallium interferes with cellular uptake of iron (Seligman, Moran, Schleicher, & David Crawford, 1992) and exerts toxicity by interference with cell cycle division processes (Rasey, Nelson, & Larson, 1981). The mechanisms of gallium toxicity are not well understood, but gallium toxicity induces a specific stress protein response that is different from arsenic or indium toxicity (Aoki, Lipsky, & Fowler, 1990) and includes heme oxygenase 1 and metallothionein-2A apparently via a mechanism involving initial formation of ROS (Yang & Chitambar, 2008). It is also used as an anticancer drug for this reason. This also means that this element, as a potent modulator of important cellular protective mechanisms such as the stress protein response, would have an impact on the stress protein response via concomitant exposure to other elements such as arsenic in gallium arsenide semiconductors (Fowler, Conner, & Yamauchi, 2005, 2008).

This type of interactive elemental information at a basic science level should be incorporated into all risk assessment analyses for semiconductor compounds containing these elements as discussed further below.

1.6 Indium

Indium is another toxic element that is used in a variety of high-speed electronic devices such as cell phones (Silveira, Fuchs, Pinheiro, Tanabe, & Bertuol, 2015), computers (Virolainen, Ibana, & Paatero, 2011), solar cells (Hau, Yip, Zou, & Jen, 2009), and flat panel televisions (Yang, Retegan, & Ekberg, 2013). It is commonly employed in common with arsenic as indium arsenide (Milnes & Polyakov, 1993) or phosphorous as indium phosphide (Metzger, 1996). More recently, gallium indium (GaIn) liquid crystal alloys have permitted the development of soft stretchable electronics (Majidi, Kramer, & Wood, 2011; Tabatabai, Fassler, Usiak, & Majidi, 2013). The production of indium for electronic devices has increased greatly in the past decades and can be expected to increase as it is used in more types of electronic devices. This element is also highly toxic and capable of inhibiting protein synthesis (Aoki et al., 1990) via a mechanism linked to degranulation of the rough endoplasmic reticulum (Fowler, Kardish, & Woods, 1983) and induction of heme oxygenase (HO-1) (Woods, Carver, & Fowler, 1979). As with gallium toxicity noted above, such a compromise protein synthesis exacerbates the toxicity of arsenic or phosphorous by attenuating cellular defense mechanisms against ROS-induced damage to important cellular machinery (Fowler et al., 2005, 2008). Lung disease has also been reported (Tanaka et al., 2010) in workers producing indium phosphide–based flat panel televisions indicating the potential risk of this disorder in persons recycling these devices under less-than-safe work facilities. Indium has been classified as a probable human carcinogen (2A) by IARC (IARC, 2006). NTP chronic inhalation studies (Program, 2001) have reported an increased incidence of lung tumors in both male and female rats and mice. Other studies (Nagano et al., 2011) have also reported an increased incidence of lung tumors in male and female rats but not in mice although clear evidence of pulmonary disease was observed in both species.

1.7 Semiconductor Compounds
1.7.1 III–V Semiconductors
As noted above, a number of electronic devices utilize combinations of gallium, arsenic, and indium as III–V semiconductors to achieve more rapid electronic flows. Combinations of these elements are light emitting and are used to produce LEDs, which are used in a variety of common devices such as clock radios and instrumentation dials (Fowler & Sexton, 2015). It is important to note that respirable particles of such semiconductor compounds will undergo biological attack in the in vivo releasing gallium (Yamauchi, Takahashi, & Yamamura, 1986), indium (Yamauchi, Takahashi, Yamamura, & Fowler, 1992), and arsenic

components. These elements are transported to distant tissues from the site of entry such as the lungs. The arsenic moiety is handled in a manner similar to As^{3+} and excreted in the urine as methylated species (Yamauchi et al., 1986, 1992) following dissolution of the GaAs or InAs moiety.

1.7.1.1 Gallium Arsenide

GaAs, which is used in a variety of instruments including computers, cell phones, and LEDs, is the most well-studied III–V semiconductor with extensive in vivo animal (Goering, Maronpot, & Fowler, 1988; Program, 2000; Tanaka, 2004; Webb, Wilson, & Carter, 1987) and in vitro study data (Burns, Sikorski, Saady, & Munson, 1991; Bustamante, Dock, Vahter, Fowler, & Orrenius, 1997; Sikorski, Burns, Stern, Luster, & Munson, 1991; Webb, Sipes, & Carter, 1984). The overall set of toxic effects seems to be a sum of both chemical toxicities from the Ga and As components following particle degradation in vivo and physical particulate effects that arise from exposure to GaAs particles themselves (Goering et al., 1988). Formation of ROS appears to be an important element in the toxicity of GaAs (Flora, Bhatt, & Mehta, 2009) (Fig 3.1).

1.7.1.2 Indium Arsenide

InAs is also a III–V semiconductor used in a variety of instruments but has a more limited database for both in vitro and in vivo toxicity studies (Bustamante et al., 1997; Conner, Yamauchi, & Fowler, 1995; Omura et al., 2000). Experimental animal studies have shown that particles of respirable dimensions undergo biological attack and partial dissolution in vivo (Yamauchi et al., 1992), resulting in the release of In and As moieties in a manner similar to particles of GaAs. The relative acute toxicity of InAs seems to be greater than that of GaAs on an equivalent dose basis (Fowler et al., 2005, 2008). ROS formation also seems to be a key element in the toxicity of InAs with inhibition of stress protein synthesis as an exacerbating factor adding to overt cell injury/cell death processes (Bustamante et al., 1997).

1.7.1.3 Indium Phosphide

Indium phosphide (InP) is a III–V semiconductor similar to those above and is used in the production of instruments including flat panel television screens and solar cells (Li, Wanlass, Gessert, Emery, & Coutts, 1989). Interstitial lung disease has been reported in workers in plants manufacturing such devices and associated with serum indium concentrations (Chonan, Taguchi, & Omae, 2007; Cummings et al., 2010). In Japanese workers, these lung effects were subsequently reported to occur in a dose-related manner with serum indium concentrations (Nakano et al., 2009). This material is highly toxic and has also been classified as a probable human carcinogen by IARC (2006) on the basis of in vivo animal (Program, 2001) and in vitro cellular studies (Bustamante et al., 1997; Tanaka et al., 1996). The mechanisms of toxicity are also linked to induction of cell death pathways such as apoptosis (Bustamante et al., 1997).

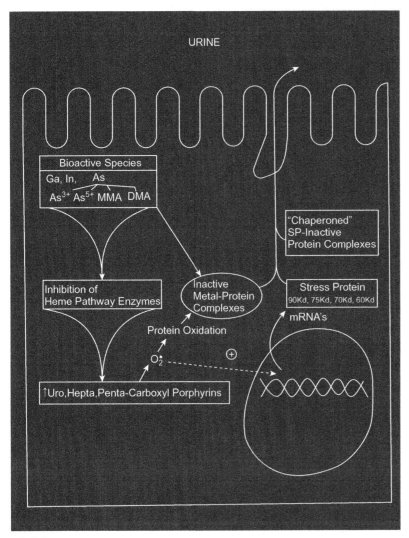

FIGURE 3.1

A global graphic from Fowler et al. (2005) showing intracellular handling and putative mechanisms of toxicity for galium (Ga), indium (In), and arsenic (As) species in a renal tubule cell following release from GaAs or InAs particles. See Fowler et al. (2005, 2008) for details.

1.7.2 II–VI Semiconductors

The II–VI semiconductors represent a second major class of semiconductors with a wide variety of applications in the electronic industry (Afzaal & O'Brien, 2006). Major representatives of this group include cadmium selenide (CdSe), cadmium sulfide (CdS), and cadmium telluride (CdTe). The toxicology database on these materials is limited, but it is reasonable to assume that particles of these materials are handled in vivo in a manner similar to that of the III–V semiconductors with both particle effects and biological attack releasing Cd^{2+} and Se and S and Te moieties; however, only limited data are available (Kirchner et al., 2005; Wang, Nagesha, Selvarasah, Dokmeci, & Carrier, 2008). In addition, nanoparticles of II–VI semiconductors including CdSe (Sun, Marx, & Greenham, 2003), CdS (Pardo-Yissar, Katz, Wasserman, & Willner, 2003), and CdTe (Kumar & Nann, 2004), among others, have also been formulated for a variety of new technological electronic devices such as solar cells (Kumar & Nann, 2004) that will eventually find their way into the e-waste stream.

1.7.2.1 Cadmium Selenide

Cadmium selenide (CdSe) is a major representative of the II–VI semiconductor group. This material is used in a variety of electronic devices including opto-electronic devices such as blue-green emitters (LEDs), solar cells (Lee, Huang, & Chien, 2008), and infrared (IR) detectors (Li et al., 2005; Nozik et al., 2010; Steckel et al., 2006; Yan, Dadvand, Rosei, & Perepichka, 2010; Zhong, Zhou, Yang, Yang, & Li, 2007). The advent of CdSe nanomaterials further expands the possible uses of these binary compounds in more miniature electronic devices, which will invariably become part the e-waste stream.

1.7.2.2 Cadmium Sulfide

Cadmium sulfide (CdS), known as cadmium yellow, is a bright yellow pigment used in paints and printer inks (Ingrosso et al., 2009; Marjanovic et al., 2011). It has a number of optoelectronic applications (Agarwal & Lieber, 2006; Li et al., 2013). As with CdSe, it is also being incorporated into nanomaterials and hence into electronic devices that will enter the e-waste stream.

1.7.2.3 Cadmium Telluride

Cadmium telluride (CdTe) is mainly used in solar cells but also finds application in IR detectors, radiation detectors, electrooptic modulators (Limousin, 2003; Singh et al., 2004; Su et al., 2010). The recycling of electronic devices such as solar panels whose use is expanding in the global move toward "green energy production" will mean these materials will be entering the e-waste stream in greater quantities in coming decades (Green, Emery, Hishikawa, Warta, & Dunlop, 2015; Sites & Pan, 2007). CdTe nanoparticles are regarded as highly toxic (Cho et al., 2007; Zhang et al., 2007), and the mechanisms of toxicity seem to be linked to the physical properties of the particles and both the Cd and Te components (Cho et al., 2007; Su et al., 2010; Yan et al., 2011). A reasonable concern is that release of CdTe nanoparticles from solid-state materials such as solar panels that could occur during recycling could result in human exposures and subsequent toxicity. The development of nanomaterial forms of CdTe can hence be expected to increase bioavailability in both environmental and occupational exposure terms.

1.7.2.4 Chromium

This metal is found in floppy disks and CDs coated with chromium dioxide (Bhushan, Theunissen, & Li, 1997) and released during the recycling process by shredding or incineration. Chromate (Cr^{6+}) has been identified as a known human carcinogen by the IARC (Boffetta, 1993).

2. ORGANIC CHEMICALS

In addition to a number of toxic metallic compounds that produce both conventional and nanomaterial exposures, electronic devices also contain a number of organic material components. Some of these materials may be released during the recycling process directly into soils in landfills or released into water bodies by runoff from landfills or by aersol exposures from incineration of plastic housings, insulation, or wire coatings (Leung, Cai, & Wong, 2006; Leung, Luksemburg, Wong, & Wong, 2007; Wong et al., 2007). Aerosolized chemicals may also be deposited over wide areas as a result of dry deposition or precipitation with rain (Tian et al., 2011; Zhang, Guan, Li, & Zeng, 2009). Some of the persistent chemicals such as the polychlorinated biphenyls (PCBs), polybrominated diphenyl ethers (PBDEs) (Luo, Cai, & Wong, 2007), and polybrominated biphenyls (PBBs) may hence also accumulate in house dust (Wang et al., 2010), fish (Wu et al., 2008), and food crops (Liu et al., 2008; Zhao et al., 2009) in areas impacted by these processes. It is also important to note that these organic chemicals frequently occur as chemical mixtures (Frazzoli, Orisakwe, Dragone, & Mantovani, 2010; Robinson, 2009; Tsydenova & Bengtsson, 2011) both with other organic compounds and the metallic compounds noted above. These combined mixture exposures and employment of child labor greatly complicate risk assessments for chemicals released from e-waste materials (Chan et al., 2007; Wang, Y., Luo, C. L., et al., 2011; Zhang et al., 2014; Zheng et al., 2013). The following discussion is a brief overview of some of the known representative types of organic chemicals associated with e-waste. This is likely not an all-inclusive list but will hopefully give the reader a sense of the general problem area and research needs going forward.

2.1 Styrene

Many electronic devices have plastic components such as composite housings, plastic keyboards, and circuit board frames. Ideally, these plastics may be broken up and recycled into new devices, but frequently this does occur efficiently, and these components may be discarded into landfills or incinerated (Wong et al., 2007). This is frequently the case in developing countries (Nnorom & Osibanjo, 2008). Styrene (acrylonitrile butadiene styrene and high-impact polystyrene) is among the major plastics in e-waste (Brennan, Isaac, & Arnold, 2002) with known toxicity and carcinogenic potential to organs such as the liver (Morgan et al., 1993) and respiratory tract (Cruzan et al., 2002). Effects on reproductive organs such as the testicular cells have also been reported (Bjørge et al., 1996). This chemical is metabolized via the cytochrome P-450 system (Kim et al., 1997), and toxicity is mediated in part by further Phase II metabolism and conjugation with glutathione (Uusküla, Järventaus, Hirvonen, Sorsa, & Norppa, 1995).

2.2 Bisphenol A

Bisphenol A is another common organic chemical found in e-waste (Huang, Zhao, Liu, & Sun, 2014; Wang & Xu, 2014), which may also persist for prolonged periods in the environment (Huang et al., 2014). It is a known endocrine-disrupting chemical (Rubin, 2011), which has been associated with both reproductive disorders (Kandaraki et al., 2010; Takeuchi, Tsutsumi, Ikezuki, Takai, & Taketani, 2004) and altered glucose regulation (Ropero et al., 2008) and development of type II diabetes (Alonso-Magdalena, Quesada, & Nadal, 2011; Magliano & Lyons, 2012; Sabanayagam, Teppala, & Shankar, 2013). This chemical is of major public health concern (Takayanagi et al., 2006) since it appears to be able to produce effects at relatively low exposure levels (Quesada et al., 2002) via interaction with hormonal receptors (Alonso-Magdalena, Morimoto, Ripoll, Fuentes, & Nadal, 2006). As discussed below, other endocrine-disrupting chemicals associated with e-waste (PCBs and dioxins) may also produce similar effects on glucose regulation leading to obesity and type II diabetes via interaction with the endocrine system (Ruiz, Perlina, Mumtaz, & Fowler, 2016).

2.3 Polychlorinated Biphenyls

PCBs are also known to be present in e-waste (Liu et al., 2008) and to be found at elevated concentrations in air, water, soils, plants, various types of foods, birds, and local residents living near the recycling facility (Luo et al., 2008) in areas near e-waste recycling sites in China (Han et al., 2010; Liu et al., 2008; Luo et al., 2011; Shen et al., 2009; Wang et al., 2012; Zhao et al., 2010). These chemicals are known to induce proliferation of smooth endoplasmic reticulum, induce cytochrome P-450 enzyme activities, and produce hepatotoxic effects (Kasza et al., 1977).

2.4 Polybrominated Biphenyls/Polybrominated Diphenyl Ethers

PBBs and PBDEs are common chemicals used as flame retardants in electronic equipment, and like many halogenated chemicals, they are persistent in the environment and may accumulate in fish (Luo et al., 2007), birds (Luo, Liu, et al., 2009), and crops (Wang, Y., Luo, C., et al., 2011) grown in soils (Luo, Luo, et al., 2009) contaminated with them from landfill or aerosol deposition (Chen et al., 2009). These chemicals produce hepatotoxic and biochemical effects similar to the PCBs as noted above (Kasza et al., 1977).

2.5 Dibenzo Dioxins and Dibenzo Furans

Dibenzo dioxins (DBDs) and dibenzo furans (DBFs) are structurally similar compounds, which have been extensively studied with regard to toxicity and carcinogenicity and as endocrine disrupting agents (Birnbaum, Staskal, & Diliberto, 2003; Van den Berg et al., 2006). Tetrachloro dibenzo dioxin (TCDD) is regarded as among the most toxic man-made chemicals known (Mukerjee, 1998). The DBDs and DBFs are known to be generated by combustion of printed circuit boards (Duan, Li, Liu, Yamazaki, & Jiang, 2011) and removal of polyvinyl chloride insulation from copper wiring by open-pit burning (Man, Naidu, & Wong, 2013; Ren, Tang, Peng, & Cai, 2015).

2.6 Chemical Mixtures and Incineration of Combustion Products

It should be noted that the above short list of chemicals is not inclusive and represents only some of the major toxic organic agents known to be present in the e-waste stream. This list of chemicals does, however, illustrate the need for considering the issue of chemical mixtures in performing risk assessments on e-waste recycling sites since exposure to these chemicals as mixtures is the most common scenario. In addition, there is need to consider combustion products of these agents from open-pit burning. Presently, the number and types of chemical combustion products that would be generated during incineration of e-waste, with the exception of TCDD and BPA, are presently poorly characterized. This is an area of much needed research since it is possible that these incineration by-products (e.g., TCDD and BPA) are also highly toxic and/or carcinogenic. Exposures of persons tending open-pit burn sites or in local communities is a cause for concern. This is particularly true for children who may experience such chemical exposures early in life and develop chronic diseases or cancer as they become adults (Birnbaum & Fenton, 2003). Hazard characterization is an essential first step in conducting a credible risk assessment and of particular importance in dealing with a complex situation detailed in this book for e-waste.

References

Adachi, S. (2009). In *Properties of semiconductor alloys: Group-IV, III-V and II-VI semiconductors* (Vol. 28). John Wiley & Sons.

Afzaal, M., & O'Brien, P. (2006). Recent developments in II–VI and III–VI semiconductors and their applications in solar cells. *Journal of Materials Chemistry, 16*(17), 1597–1602.

Agarwal, R., & Lieber, C. (2006). Semiconductor nanowires: Optics and optoelectronics. *Applied Physics A, 85*(3), 209–215.

Akinloye, O., Arowojolu, A. O., Shittu, O. B., & Anetor, J. I. (2006). Cadmium toxicity: A possible cause of male infertility in Nigeria. *Reproductive Biology, 6*(1), 17–30.

Alonso-Magdalena, P., Morimoto, S., Ripoll, C., Fuentes, E., & Nadal, A. (2006). The estrogenic effect of bisphenol A disrupts pancreatic β-cell function in vivo and induces insulin resistance. *Environmental Health Perspectives*, 106–112.

Alonso-Magdalena, P., Quesada, I., & Nadal, A. (2011). Endocrine disruptors in the etiology of type 2 diabetes mellitus. *Nature Reviews Endocrinology, 7*(6), 346–353.

Aoki, Y., Lipsky, M. M., & Fowler, B. A. (1990). Alteration in protein synthesis in primary cultures of rat kidney proximal tubule epithelial cells by exposure to gallium, indium, and arsenite. *Toxicology and Applied Pharmacology, 106*(3), 462–468.

Aposhian, H. V., Zakharyan, R. A., Avram, M. D., Sampayo-Reyes, A., & Wollenberg, M. L. (2004). A review of the enzymology of arsenic metabolism and a new potential role of hydrogen peroxide in the detoxication of the trivalent arsenic species. *Toxicology and Applied Pharmacology, 198*(3), 327–335.

Babu, B. R., Parande, A. K., & Basha, C. A. (2007). Electrical and electronic waste: A global environmental problem. *Waste Management and Research, 25*(4), 307–318.

Betts, K. S. (2011). A study in balance: How microbiomes are changing the shape of environmental health. *Environmental Health Perspectives, 119*(8), A340–A346.

Bhattacharyya, M. H. (1991). Cadmium-induced bone loss: Increased susceptibility in females. *Water, Air, and Soil Pollution, 57*(1), 665–673.

Bhushan, B., Theunissen, G. S., & Li, X. (1997). Tribological studies of chromium oxide films for magnetic recording applications. *Thin Solid Films, 311*(1), 67–80.

Birnbaum, L. S., & Fenton, S. E. (2003). Cancer and developmental exposure to endocrine disruptors. *Environmental Health Perspectives, 111*(4), 389.

Birnbaum, L. S., Staskal, D. F., & Diliberto, J. J. (2003). Health effects of polybrominated dibenzo-p-dioxins (PBDDs) and dibenzofurans (PBDFs). *Environment International, 29*(6), 855–860.

Bjørge, C., Brunborg, G., Wiger, R., Holme, J. A., Scholz, T., Dybing, E., & Søderlund, E. J. (1996). A comparative study of chemically induced DNA damage in isolated human and rat testicular cells. *Reproductive Toxicology, 10*(6), 509–519.

Boffetta, P. (1993). Carcinogenicity of trace elements with reference to evaluations made by the International Agency for Research on Cancer. *Scandinavian Journal of Work, Environment and Health,* 67–70.

Bolt, A. M., Zhao, F., Pacheco, S., & Klimecki, W. T. (2012). Arsenite-induced autophagy is associated with proteotoxicity in human lymphoblastoid cells. *Toxicology and Applied Pharmacology, 264*(2), 255–261.

Brennan, L., Isaac, D., & Arnold, J. (2002). Recycling of acrylonitrile–butadiene–styrene and high-impact polystyrene from waste computer equipment. *Journal of Applied Polymer Science, 86*(3), 572–578.

Burns, L., Sikorski, E., Saady, J., & Munson, A. (1991). Evidence for arsenic as the immunosuppressive component of gallium arsenide. *Toxicology and Applied Pharmacology, 110*(1), 157–169.

Bustamante, J., Dock, L., Vahter, M., Fowler, B., & Orrenius, S. (1997). The semiconductor elements arsenic and indium induce apoptosis in rat thymocytes. *Toxicology, 118*(2), 129–136.

Bustamante, J., Nutt, L., Orrenius, S., & Gogvadze, V. (2005). Arsenic stimulates release of cytochrome c from isolated mitochondria via induction of mitochondrial permeability transition. *Toxicology and Applied Pharmacology, 207*(2), 110–116.

Chan, J. K., Xing, G. H., Xu, Y., Liang, Y., Chen, L. X., Wu, S. C., … Wong, M. H. (2007). Body loadings and health risk assessment of polychlorinated dibenzo-p-dioxins and dibenzofurans at an intensive electronic waste recycling site in China. *Environmental Science and Technology, 41*(22), 7668–7674.

Chedrese, P. J., Piasek, M., & Henson, M. C. (2006). Cadmium as an endocrine disruptor in the reproductive system. *Immunology, Endocrine and Metabolic Agents in Medicinal Chemistry (Formerly Current Medicinal Chemistry-Immunology, Endocrine and Metabolic Agents), 6*(1), 27–35.

Chen, D., Bi, X., Zhao, J., Chen, L., Tan, J., Mai, B., … Wong, M. (2009). Pollution characterization and diurnal variation of PBDEs in the atmosphere of an E-waste dismantling region. *Environmental Pollution, 157*(3), 1051–1057.

Cherian, M. G., Jayasurya, A., & Bay, B.-H. (2003). Metallothioneins in human tumors and potential roles in carcinogenesis. *Mutation Research/Fundamental and Molecular Mechanisms of Mutagenesis, 533*(1), 201–209.

Cho, S. J., Maysinger, D., Jain, M., Röder, B., Hackbarth, S., & Winnik, F. M. (2007). Long-term exposure to CdTe quantum dots causes functional impairments in live cells. *Langmuir, 23*(4), 1974–1980.

Chonan, T., Taguchi, O., & Omae, K. (2007). Interstitial pulmonary disorders in indium-processing workers. *European Respiratory Journal, 29*(2), 317–324.

Clarkson, T. W., & Magos, L. (2006). The toxicology of mercury and its chemical compounds. *Critical Reviews in Toxicology, 36*(8), 609–662.

Clarkson, T. W., Vyas, J. B., & Ballatori, N. (2007). Mechanisms of mercury disposition in the body. *American Journal of Industrial Medicine, 50*(10), 757–764.

Conner, E. A., Yamauchi, H., & Fowler, B. A. (1995). Alterations in the heme biosynthetic pathway from the III-V semiconductor metal, indium arsenide (InAs). *Chemico-Biological Interactions, 96*(3), 273–285.

Cruzan, G., Carlson, G. P., Johnson, K. A., Andrews, L. S., Banton, M. I., Bevan, C., & Cushman, J. R. (2002). Styrene respiratory tract toxicity and mouse lung tumors are mediated by CYP2F-generated metabolites. *Regulatory Toxicology and Pharmacology, 35*(3), 308–319.

Cummings, K. J., Donat, W. E., Ettensohn, D. B., Roggli, V. L., Ingram, P., & Kreiss, K. (2010). Pulmonary alveolar proteinosis in workers at an indium processing facility. *American Journal of Respiratory and Critical Care Medicine, 181*(5), 458–464.

Dolzhnikov, D. S., Zhang, H., Jang, J., Son, J. S., Panthani, M. G., Shibata, T., … Talapin, D. V. (2015). Composition-matched molecular "solders" for semiconductors. *Science, 347*(6220), 425–428.

Duan, H., Li, J., Liu, Y., Yamazaki, N., & Jiang, W. (2011). Characterization and inventory of PCDD/Fs and PBDD/Fs emissions from the incineration of waste printed circuit board. *Environmental Science and Technology, 45*(15), 6322–6328.

Fawcett, S. E., & Jamieson, H. E. (2011). The distinction between ore processing and post-depositional transformation on the speciation of arsenic and antimony in mine waste and sediment. *Chemical Geology, 283*(3), 109–118.

Flora, S. J. (2011). Arsenic-induced oxidative stress and its reversibility. *Free Radical Biology and Medicine, 51*(2), 257–281.

Flora, S. J., Bhatt, K., & Mehta, A. (2009). Arsenic moiety in gallium arsenide is responsible for neuronal apoptosis and behavioral alterations in rats. *Toxicology and Applied Pharmacology, 240*(2), 236–244.

Fowler, B. A. (1993). *NAS/NRC. Report of the committee on measuring lead exposure in infants, children and other sensitive populations* (pp. 337). Washington, D.C: NAS/NRC Press.

Fowler, B. A. (1998). Roles of lead-binding proteins in mediating lead bioavailability. *Environmental Health Perspectives, 106*(Suppl. 6), 1585–1587.

Fowler, B. A. (2009). Monitoring of human populations for early markers of cadmium toxicity: A review. *Toxicology and Applied Pharmacology, 238*(3), 294–300.

Fowler, B. A. (Ed.). (2013a). *Computational toxicology: Applications for risk assessment* (pp. 258). Amsterdam: Elsevier Publishers.

Fowler, B. A. (2013b). Cadmium and aging. In B. Weiss (Ed.), *Aging and vulnerabilities to environmental chemicals. Royal society of chemistry* (pp. 376–387). UK: Cambridge.

Fowler, B. A., Conner, E. A., & Yamauchi, H. (2005). Metabolomic and proteomic biomarkers for III-V semiconductors: Chemical-specific porphyrinurias and proteinurias. *Toxicology and Applied Pharmacology, 206*(2), 121–130. http://dx.doi.org/10.1016/j.taap.2005.01.020.

Fowler, B. A., Conner, E. A., & Yamauchi, H. (2008). Proteomic and metabolomic biomarkers for III-V semiconductors: prospects for applications to nano-materials. *Toxicology and Applied Pharmacology, 233*(1), 110–115.

Fowler, B. A., Kardish, R., & Woods, J. S. (1983). Alteration of hepatic microsomal structure and function by acute indium administration: Ultrastructural morphometric and biochemical studies. *Laboratory Investigation, 48*, 471–478.

Fowler, B. A., & Sexton, M. J. (2015). Gallium and semiconductor compounds. In G. F. Nordberg, B. A. Fowler, & M. Nordberg (Eds.), *Handbook on the toxicology of metals* (4th ed.) (pp. 787–797). Amsterdam: Elsevier Publishers.

Fowler, B. A., & Woods, J. S. (1977). Ultrastructural and biochemical changes in renal mitochondria during chronic oral methyl mercury exposure: The relationship to renal function. *Experimental and Molecular Pathology, 27*(3), 403–412.

Frazzoli, C., Orisakwe, O. E., Dragone, R., & Mantovani, A. (2010). Diagnostic health risk assessment of electronic waste on the general population in developing countries' scenarios. *Environmental Impact Assessment Review, 30*(6), 388–399.

García-Hernández, J., Cadena-Cárdenas, L., Betancourt-Lozano, M., García-De-La-Parra, L. M., García-Rico, L., & Márquez-Farías, F. (2007). Total mercury content found in edible tissues of top predator fish from the Gulf of California, Mexico. *Toxicological and Environmental Chemistry, 89*(3), 507–522.

Goering, P. L., Maronpot, R. R., & Fowler, B. A. (1988). Effect of intratracheal gallium arsenide administration on δ-aminolevulinic acid dehydratase in rats: Relationship to urinary excretion of aminolevulinic acid. *Toxicology and Applied Pharmacology, 92*(2), 179–193.

Green, M. A., Emery, K., Hishikawa, Y., Warta, W., & Dunlop, E. D. (2015). Solar cell efficiency tables (Version 45). *Progress in Photovoltaics: Research and Applications, 23*(1), 1–9.

Han, W., Feng, J., Gu, Z., Wu, M., Sheng, G., & Fu, J. (2010). Polychlorinated biphenyls in the atmosphere of Taizhou, a major e-waste dismantling area in China. *Journal of Environmental Sciences, 22*(4), 589–597.

Hau, S. K., Yip, H.-L., Zou, J., & Jen, A. K. -Y. (2009). Indium tin oxide-free semi-transparent inverted polymer solar cells using conducting polymer as both bottom and top electrodes. *Organic Electronics, 10*(7), 1401–1407.

Henke, K. (2009). *Arsenic: Environmental chemistry, health threats and waste treatment.* John Wiley & Sons.

Herat, S. (2008). Recycling of cathode ray tubes (CRTs) in electronic waste. *CLEAN–Soil, Air, Water, 36*(1), 19–24.

Hightower, J. M., & Moore, D. (2003). Mercury levels in high-end consumers of fish. *Environmental Health Perspectives, 111*(4), 604.

Hsu, P.-C., & Guo, Y. L. (2002). Antioxidant nutrients and lead toxicity. *Toxicology, 180*(1), 33–44.

Huang, D.-Y., Zhao, H.-Q., Liu, C.-P., & Sun, C.-X. (2014). Characteristics, sources, and transport of tetrabromobisphenol A and bisphenol A in soils from a typical e-waste recycling area in South China. *Environmental Science and Pollution Research, 21*(9), 5818–5826.

IARC. (2006). *IARC monographs on the evaluation of carcinogenic risks to humans. Cobalt in hard metals, and cobalt sulfate, gallium arsenide, indium phosphide and vanadium pentoxide* (pp. 330). Lyon France.

Ingrosso, C., Kim, J. Y., Binetti, E., Fakhfouri, V., Striccoli, M., Agostiano, A., ... Brugger, J. (2009). Drop-on-demand inkjet printing of highly luminescent CdS and CdSe@ZnS nanocrystal based nanocomposites. *Microelectronic Engineering, 86*(4), 1124–1126.

Kandaraki, E., Chatzigeorgiou, A., Livadas, S., Palioura, E., Economou, F., Koutsilieris, M., ... Diamanti-Kandarakis, E. (2010). Endocrine disruptors and polycystic ovary syndrome (PCOS): Elevated serum levels of bisphenol A in women with PCOS. *The Journal of Clinical Endocrinology and Metabolism*, *96*(3), E480–E484.

Kasza, L., Weinberger, M., Hinton, D., Trump, B., Patel, C., Friedman, L., & Garthoff, L. (1977). Comparative toxicity of polychlorinated biphenyl and polybrominated biphenyl in the rat liver: Light and electron microscopic alterations after subacute dietary exposure. *Journal of Environmental Pathology and Toxicology*, *1*(3), 241–257.

Kazantzis, G. (2004). Cadmium, osteoporosis and calcium metabolism. *Biometals*, *17*(5), 493–498.

Kim, H., Wang, R., Elovaara, E., Raunio, H., Pelkonen, O., Aoyama, T., ... Nakajima, T. (1997). Cytochrome P450 isozymes responsible for the metabolism of toluene and styrene in human liver microsomes. *Xenobiotica*, *27*(7), 657–665.

Kirchner, C., Liedl, T., Kudera, S., Pellegrino, T., Muñoz Javier, A., Gaub, H. E., ... Parak, W. J. (2005). Cytotoxicity of colloidal CdSe and CdSe/ZnS nanoparticles. *Nano Letters*, *5*(2), 331–338.

Klaassen, C. D., Liu, J., & Diwan, B. A. (2009). Metallothionein protection of cadmium toxicity. *Toxicology and Applied Pharmacology*, *238*(3), 215–220.

Kumar, S., & Nann, T. (2004). First solar cells based on CdTe nanoparticle/MEH-PPV composites. *Journal of Materials Research*, *19*(07), 1990–1994.

Lee, Y.-L., Huang, B.-M., & Chien, H.-T. (2008). Highly efficient CdSe-sensitized TiO_2 photo-electrode for quantum-dot-sensitized solar cell applications. *Chemistry of Materials*, *20*(22), 6903–6905.

Leung, A., Cai, Z. W., & Wong, M. H. (2006). Environmental contamination from electronic waste recycling at Guiyu, southeast China. *Journal of Material Cycles and Waste Management*, *8*(1), 21–33.

Leung, A. O., Luksemburg, W. J., Wong, A. S., & Wong, M. H. (2007). Spatial distribution of poly-brominated diphenyl ethers and polychlorinated dibenzo-p-dioxins and dibenzofurans in soil and combusted residue at Guiyu, an electronic waste recycling site in southeast China. *Environmental Science and Technology*, *41*(8), 2730–2737.

Li, J., Lu, H., Guo, J., Xu, Z., & Zhou, Y. (2007). Recycle technology for recovering resources and products from waste printed circuit boards. *Environmental Science and Technology*, *41*(6), 1995–2000.

Li, Y., Rizzo, A., Mazzeo, M., Carbone, L., Manna, L., Cingolani, R., & Gigli, G. (2005). White organic light-emitting devices with CdSe/ZnS quantum dots as a red emitter. *Journal of Applied Physics*, *97*(11), 113501.

Li, H., Wang, X., Xu, J., Zhang, Q., Bando, Y., Golberg, D., ... Zhai, T. (2013). One-dimensional CdS nanostructures: A promising candidate for optoelectronics. *Advanced Materials*, *25*(22), 3017–3037.

Li, X., Wanlass, M., Gessert, T., Emery, K., & Coutts, T. (1989). High-efficiency indium tin oxide/indium phosphide solar cells. *Applied Physics Letters*, *54*(26), 2674–2676.

Lim, S.-R., & Schoenung, J. M. (2010). Human health and ecological toxicity potentials due to heavy metal content in waste electronic devices with flat panel displays. *Journal of Hazardous Materials*, *177*(1), 251–259.

Limousin, O. (2003). New trends in CdTe and CdZnTe detectors for X and gamma-ray applications. *Nuclear Instruments and Methods in Physics Research Section A: Accelerators, Spectrometers, Detectors and Associated Equipment*, *504*(1), 24–37.

Liu, H., Zhou, Q., Wang, Y., Zhang, Q., Cai, Z., & Jiang, G. (2008). E-waste recycling induced poly-brominated diphenyl ethers, polychlorinated biphenyls, polychlorinated dibenzo-p-dioxins and dibenzo-furans pollution in the ambient environment. *Environment International*, *34*(1), 67–72.

Lund, B.-O., Miller, D. M., & Woods, J. S. (1991). Mercury-induced H_2O_2 production and lipid peroxidation in vitro in rat kidney mitochondria. *Biochemical Pharmacology, 42*, S181–S187.

Luo, Q., Cai, Z. W., & Wong, M. H. (2007). Polybrominated diphenyl ethers in fish and sediment from river polluted by electronic waste. *Science of the Total Environment, 383*(1), 115–127.

Luo, X.-J., Liu, J., Luo, Y., Zhang, X.-L., Wu, J.-P., Lin, Z., … Yang, Z.-Y. (2009). Polybrominated diphenyl ethers (PBDEs) in free-range domestic fowl from an e-waste recycling site in South China: Levels, profile and human dietary exposure. *Environment International, 35*(2), 253–258.

Luo, C., Liu, C., Wang, Y., Liu, X., Li, F., Zhang, G., & Li, X. (2011). Heavy metal contamination in soils and vegetables near an e-waste processing site, south China. *Journal of Hazardous Materials, 186*(1), 481–490.

Luo, Y., Luo, X.-J., Lin, Z., Chen, S.-J., Liu, J., Mai, B.-X., & Yang, Z.-Y. (2009). Polybrominated diphenyl ethers in road and farmland soils from an e-waste recycling region in Southern China: Concentrations, source profiles, and potential dispersion and deposition. *Science of the Total Environment, 407*(3), 1105–1113.

Luo, X.-J., Zhang, X.-L., Liu, J., Wu, J.-P., Luo, Y., Chen, S.-J., … Yang, Z.-Y. (2008). Persistent halogenated compounds in waterbirds from an e-waste recycling region in South China. *Environmental Science and Technology, 43*(2), 306–311.

Magliano, D. J., & Lyons, J. G. (2012). Bisphenol A and diabetes, insulin resistance, cardiovascular disease and obesity: Controversy in a (plastic) cup? *The Journal of Clinical Endocrinology and Metabolism, 98*(2), 502–504.

Majidi, C., Kramer, R., & Wood, R. (2011). A non-differential elastomer curvature sensor for softer-than-skin electronics. *Smart Materials and Structures, 20*(10), 105017.

Man, M., Naidu, R., & Wong, M. H. (2013). Persistent toxic substances released from uncontrolled e-waste recycling and actions for the future. *Science of the Total Environment, 463*, 1133–1137.

Marjanovic, N., Hammerschmidt, J., Perelaer, J., Farnsworth, S., Rawson, I., Kus, M., … Baumann, R. R. (2011). Inkjet printing and low temperature sintering of CuO and CdS as functional electronic layers and Schottky diodes. *Journal of Materials Chemistry, 21*(35), 13634–13639.

Metzger, R. (1996). The capabilities of indium phosphide electronics and optoelectronics. *Compound Semiconductor, 2*(2), 20–24.

Meyer, P. A., Brown, M. J., & Falk, H. (2008). Global approach to reducing lead exposure and poisoning. *Mutation Research/Reviews in Mutation Research, 659*(1), 166–175.

Milnes, A., & Polyakov, A. (1993). Indium arsenide: A semiconductor for high speed and electro-optical devices. *Materials Science and Engineering: B, 18*(3), 237–259.

Morgan, D., Mahler, J., Dill, J., Price, H., O'Cornnor, R., & Adkins, B. (1993). Styrene inhalation toxicity studies in mice III. Strain differences in susceptibility. *Toxicological Sciences, 21*(3), 326–333.

Moskalyk, R. (2003). Gallium: The backbone of the electronics industry. *Minerals Engineering, 16*(10), 921–929.

Mukerjee, D. (1998). Health impact of polychlorinated dibenzo-p-dioxins: A critical review. *Journal of the Air and Waste Management Association, 48*(2), 157–165.

Nagano, K., Nishizawa, T., Umeda, Y., Kasai, T., Noguchi, T., Gotoh, K., … Yamauchi, T. (2011). Inhalation carcinogenicity and chronic toxicity of indium-tin oxide in rats and mice. *Journal of Occupational Health, 53*(3), 175–187.

Nakada, S., Nomoto, A., & Imura, N. (1980). Effect of methylmercury and inorganic mercury on protein synthesis in mammalian cells. *Ecotoxicology and Environmental Safety, 4*(2), 184–190.

Nakano, M., Omae, K., Tanaka, A., Hirata, M., Michikawa, T., Kikuchi, Y., … Chonan, T. (2009). Causal relationship between indium compound inhalation and effects on the lungs. *Journal of Occupational Health, 51*(6), 513–521.

Navas-Acien, A., Guallar, E., Silbergeld, E. K., & Rothenberg, S. J. (2007). Lead exposure and cardiovascular disease: A systematic review. *Environmental Health Perspectives*, 472–482.

Needleman, H. (2004). Lead poisoning. *Annual Review of Medicine*, 55, 209–222.

Nishijo, M., Satarug, S., Honda, R., Tsuritani, I., & Aoshima, K. (2004). The gender differences in health effects of environmental cadmium exposure and potential mechanisms. *Molecular and Cellular Biochemistry*, 255(1–2), 87–92.

Nnorom, I. C., & Osibanjo, O. (2008). Overview of electronic waste (e-waste) management practices and legislations, and their poor applications in the developing countries. *Resources, Conservation and Recycling*, 52(6), 843–858.

Nordberg, G. F., Fowler, B. A., & Nordberg, M. (2015). *Handbook on the toxicology of metals* (4th ed.), 1385. Amsterdam: Elsevier Publishers.

Norseth, T., & Brendeford, M. (1971). Intracellular distribution of inorganic and organic mercury in rat liver after exposure to methylmercury salts. *Biochemical Pharmacology*, 20(6), 1101–1107.

Nozik, A. J., Beard, M. C., Luther, J. M., Law, M., Ellingson, R. J., & Johnson, J. C. (2010). Semiconductor quantum dots and quantum dot arrays and applications of multiple exciton generation to third-generation photovoltaic solar cells. *Chemical Reviews*, 110(11), 6873–6890.

Oken, E., Kleinman, K. P., Berland, W. E., Simon, S. R., Rich-Edwards, J. W., & Gillman, M. W. (2003). Decline in fish consumption among pregnant women after a national mercury advisory. *Obstetrics and Gynecology*, 102(2), 346.

Omura, M., Yamazaki, K., Tanaka, A., Hirata, M., Makita, Y., & Inoue, N. (2000). Changes in the testicular damage caused by indium arsenide and indium phosphide in hamsters during two years after intratracheal instillations. *Journal of Occupational Health*, 42(4), 196–204.

Onalaja, A. O., & Claudio, L. (2000). Genetic susceptibility to lead poisoning. *Environmental Health Perspectives*, 108(Suppl. 1), 23.

Oskarsson, A., & Fowler, B. A. (1985a). Effects of lead inclusion bodies on subcellular distribution of lead in rat kidney: The relationship to mitochondrial function. *Experimental and Molecular Pathology*, 43(3), 397–408.

Oskarsson, A., & Fowler, B. A. (1985b). Effects of lead on the heme biosynthetic pathway in rat kidney. *Experimental and Molecular Pathology*, 43(3), 409–417.

Pardo-Yissar, V., Katz, E., Wasserman, J., & Willner, I. (2003). Acetylcholine esterase-labeled CdS nanoparticles on electrodes: Photoelectrochemical sensing of the enzyme inhibitors. *Journal of the American Chemical Society*, 125(3), 622–623.

Parks, J. M., Johs, A., Podar, M., Bridou, R., Hurt, R. A., Smith, S. D., … Liang, L. (2013). The genetic basis for bacterial mercury methylation. *Science*, 339(6125), 1332–1335.

Podar, M., Gilmour, C. C., Brandt, C. C., Soren, A., Brown, S. D., … Crable, B. R., & Elias, D. A. (2015). Global prevalence and distribution of genes and microorganisms involved in mercury methylation. *Science Advances*, 1(9), e1500675.

Program, N. T. (2000). NTP toxicology and carcinogenesis studies of gallium arsenide (CAS No. 1303-00-0) in F344/N rats and B6C3F1 mice (inhalation studies). *National Toxicology Program Technical Report Series*, 492, 1.

Program, N. T. (2001). Toxicology and carcinogenesis studies of indium phosphide (CAS No. 22398-90-7) in F344/N rats and B6C3F1 mice (inhalation studies). *National Toxicology Program Technical Report Series*, 499, 7.

Prozialeck, W. C. (2000). Evidence that E-cadherin may be a target for cadmium toxicity in epithelial cells. *Toxicology and Applied Pharmacology*, 164(3), 231–249.

Quesada, I., Fuentes, E., Viso-León, M. C., Soria, B., Ripoll, C., & Nadal, A. (2002). Low doses of the endocrine disruptor bisphenol-A and the native hormone 17β-estradiol rapidly activate transcription factor CREB. *The FASEB Journal*, 16(12), 1671–1673.

Rajan, S., & Jena, D. (2013). Gallium nitride electronics. *Semiconductor Science and Technology,* *28*(7), 070301.

Rasey, J. S., Nelson, N. J., & Larson, S. M. (1981). Relationship of iron metabolism to tumor cell toxicity of stable gallium salts. *International Journal of Nuclear Medicine and Biology, 8*(4), 303–313.

Ren, M., Tang, Y., Peng, P., & Cai, Y. (2015). PCDD/Fs in air and soil around an e-waste dismantling area with open burning of insulated wires in south China. *Bulletin of Environmental Contamination and Toxicology, 94*(5), 647–652.

Robinson, B. H. (2009). E-waste: An assessment of global production and environmental impacts. *Science of the Total Environment, 408*(2), 183–191.

Ropero, A., Alonso-Magdalena, P., García-García, E., Ripoll, C., Fuentes, E., & Nadal, A. (2008). Bisphenol-A disruption of the endocrine pancreas and blood glucose homeostasis. *International Journal of Andrology, 31*(2), 194–200.

Rubin, B. S. (2011). Bisphenol A: An endocrine disruptor with widespread exposure and multiple effects. *The Journal of Steroid Biochemistry and Molecular Biology, 127*(1), 27–34.

Ruiz, P., Mumtaz, M., Osterloh, J., Fisher, J., & Fowler, B. A. (2010). Interpreting NHANES biomonitoring data, cadmium. *Toxicology Letters, 198*(1), 44–48.

Ruiz, P., Perlina, A., Mumtaz, M., & Fowler, B. A. (2016). A systems biology approach reveals converging molecular mechanisms that link different POPs to common metabolic diseases. *Environmental Health Perspectives, 124*(7), 1034–1041.

Sabanayagam, C., Teppala, S., & Shankar, A. (2013). Relationship between urinary bisphenol A levels and prediabetes among subjects free of diabetes. *Acta Diabetologica, 50*(4), 625–631.

Safe, S. (2003). Cadmium's disguise dupes the estrogen receptor. *Nature Medicine, 9*(8), 1000–1001.

Samikkannu, T., Chen, C.-H., Yih, L.-H., Wang, A. S., Lin, S.-Y., Chen, T.-C., & Jan, K.-Y. (2003). Reactive oxygen species are involved in arsenic trioxide inhibition of pyruvate dehydrogenase activity. *Chemical Research in Toxicology, 16*(3), 409–414.

Scinicariello, F., Yesupriya, A., Chang, M. H., & Fowler, B. A. (2010). Modification by ALAD of the association between blood lead and blood pressure in the U.S. population: Results from the Third National Health and Nutrition Examination Survey. *Environmental Health Perspectives, 118*(2), 259–264. http://dx.doi.org/10.1289/ehp.0900866.

Seligman, P. A., Moran, P. L., Schleicher, R. B., & David Crawford, E. (1992). Treatment with gallium nitrate: Evidence for interference with iron metabolism in vivo. *American Journal of Hematology, 41*(4), 232–240.

Shelton, K. R., Todd, J. M., & Egle, P. M. (1986). The induction of stress-related proteins by lead. *Journal of Biological Chemistry, 261*(4), 1935–1940.

Shen, C., Chen, Y., Huang, S., Wang, Z., Yu, C., Qiao, M., … Lin, Q. (2009). Dioxin-like compounds in agricultural soils near e-waste recycling sites from Taizhou area, China: Chemical and bioanalytical characterization. *Environment International, 35*(1), 50–55.

Shenker, B. J., Guo, T. L., & Shapiro, I. M. (1998). Low-level methylmercury exposure causes human T-cells to undergo apoptosis: Evidence of mitochondrial dysfunction. *Environmental Research, 77*(2), 149–159.

Shi, H., Hudson, L. G., & Liu, K. J. (2004). Oxidative stress and apoptosis in metal ion-induced carcinogenesis. *Free Radical Biology and Medicine, 37*(5), 582–593.

Shi, H., Shi, X., & Liu, K. J. (2004). Oxidative mechanism of arsenic toxicity and carcinogenesis. *Molecular and Cellular Biochemistry, 255*(1–2), 67–78.

Sikorski, E., Burns, L., Stern, M., Luster, M., & Munson, A. E. (1991). Splenic cell targets in gallium arsenide-induced suppression of the primary antibody response. *Toxicology and Applied Pharmacology, 110*(1), 129–142.

Silveira, A., Fuchs, M., Pinheiro, D., Tanabe, E. H., & Bertuol, D. A. (2015). Recovery of indium from LCD screens of discarded cell phones. *Waste Management, 45*, 334–342.

Singh, R., Rangari, V., Sanagapalli, S., Jayaraman, V., Mahendra, S., & Singh, V. (2004). Nano-structured CdTe, CdS and TiO_2 for thin film solar cell applications. *Solar Energy Materials and Solar Cells, 82*(1), 315–330.

Sites, J., & Pan, J. (2007). Strategies to increase CdTe solar-cell voltage. *Thin Solid Films, 515*(15), 6099–6102.

Squibb, K. S., Pritchard, J. B., & Fowler, B. A. (1984). Cadmium-Metallothionein nephropathy: Relationships between ultrastructural/biochemical alterations and intracellular cadmium binding. *Journal of Pharmacology and Experimental Therapeutics, 229*(1), 311–321.

Stanhill, A., Haynes, C. M., Zhang, Y., Min, G., Steele, M. C., Kalinina, J., … Ron, D. (2006). An arsenite-inducible 19S regulatory particle-associated protein adapts proteasomes to proteotoxicity. *Molecular Cell, 23*(6), 875–885.

Steckel, J. S., Snee, P., Coe-Sullivan, S., Zimmer, J. P., Halpert, J. E., Anikeeva, P., … Bawendi, M. G. (2006). Color-saturated green-emitting QD-LEDs. *Angewandte Chemie International Edition, 45*(35), 5796–5799.

Styblo, M., Del Razo, L. M., Vega, L., Germolec, D. R., LeCluyse, E. L., Hamilton, G. A., … Thomas, D. J. (2000). Comparative toxicity of trivalent and pentavalent inorganic and methylated arsenicals in rat and human cells. *Archives of Toxicology, 74*(6), 289–299.

Su, Y., Hu, M., Fan, C., He, Y., Li, Q., Li, W., … Huang, Q. (2010). The cytotoxicity of CdTe quantum dots and the relative contributions from released cadmium ions and nanoparticle properties. *Biomaterials, 31*(18), 4829–4834.

Suda, I., Suda, M., & Hirayama, K. (1993). Phagocytic cells as a contributor to in vivo degradation of alkyl mercury. *Bulletin of Environmental Contamination and Toxicology, 51*(3), 394–400.

Suganuma, K. (2001). Advances in lead-free electronics soldering. *Current Opinion in Solid State and Materials Science, 5*(1), 55–64.

Sun, B., Marx, E., & Greenham, N. C. (2003). Photovoltaic devices using blends of branched CdSe nanoparticles and conjugated polymers. *Nano Letters, 3*(7), 961–963.

Syversen, T. L. (1981). Effects of methyl mercury on protein synthesis in vitro. *Acta Pharmacologica et Toxicologica, 49*(5), 422–426.

Szuster-Ciesielska, A., Stachura, A., Słotwińska, M., Kamińska, T., Śnieżko, R., Paduch, R., … Kandefer-Szerszeń, M. (2000). The inhibitory effect of zinc on cadmium-induced cell apoptosis and reactive oxygen species (ROS) production in cell cultures. *Toxicology, 145*(2), 159–171.

Tabatabai, A., Fassler, A., Usiak, C., & Majidi, C. (2013). Liquid-phase gallium–indium alloy electronics with microcontact printing. *Langmuir, 29*(20), 6194–6200.

Takayanagi, S., Tokunaga, T., Liu, X., Okada, H., Matsushima, A., & Shimohigashi, Y. (2006). Endocrine disruptor bisphenol A strongly binds to human estrogen-related receptor γ (ERRγ) with high constitutive activity. *Toxicology Letters, 167*(2), 95–105.

Takebayashi, S., Jimi, S., Segawa, M., & Kiyoshi, Y. (2000). Cadmium induces osteomalacia mediated by proximal tubular atrophy and disturbances of phosphate reabsorption. A study of 11 autopsies. *Pathology-Research and Practice, 196*(9), 653–663.

Takeuchi, T., Tsutsumi, O., Ikezuki, Y., Takai, Y., & Taketani, Y. (2004). Positive relationship between androgen and the endocrine disruptor, bisphenol A, in normal women and women with ovarian dysfunction. *Endocrine Journal, 51*(2), 165–169.

Takiguchi, M., & Yoshihara, S. (2005). New aspects of cadmium as endocrine disruptor. *Environmental Sciences: An International Journal of Environmental Physiology and Toxicology, 13*(2), 107–116.

Tanaka, A. (2004). Toxicity of indium arsenide, gallium arsenide, and aluminium gallium arsenide. *Toxicology and Applied Pharmacology, 198*(3), 405–411.

Tanaka, A., Hirata, M., Kiyohara, Y., Nakano, M., Omae, K., Shiratani, M., & Koga, K. (2010). Review of pulmonary toxicity of indium compounds to animals and humans. *Thin Solid Films, 518*(11), 2934–2936.

Tanaka, A., Hisanaga, A., Hirata, M., Omura, M., Makita, Y., Inoue, N., & Ishinishi, N. (1996). Chronic toxicity of indium arsenide and indium phosphide to the lungs of hamsters. *Fukuoka Igaku Zasshi = Hukuoka Acta Medica, 87*(5), 108–115.

Tandon, S., Singh, S., Prasad, S., & Mathur, N. (2001). Hepatic and renal metallothionein induction by an oral equimolar dose of zinc, cadmium or mercury in mice. *Food and Chemical Toxicology, 39*(6), 571–577.

Tellez-Plaza, M., Navas-Acien, A., Crainiceanu, C. M., Sharrett, A. R., & Guallar, E. (2010). Cadmium and peripheral arterial disease: Gender differences in the 1999–2004 US National Health and Nutrition Examination Survey. *American Journal of Epidemiology, 172*(6), 671–681.

Thirumoorthy, N., Sunder, A. S., Kumar, K. M., Ganesh, G., & Chatterjee, M. (2011). A review of metallothionein isoforms and their role in pathophysiology. *World Journal of Surgical Oncology, 9*(1), 1.

Thomas, D. J., Li, J., Waters, S. B., Xing, W., Adair, B. M., Drobna, Z., … Styblo, M. (2007). Arsenic (+3 oxidation state) methyltransferase and the methylation of arsenicals. *Experimental Biology and Medicine, 232*(1), 3–13.

Thompson, J., & Bannigan, J. (2008). Cadmium: Toxic effects on the reproductive system and the embryo. *Reproductive Toxicology, 25*(3), 304–315.

Tian, M., Chen, S.-J., Wang, J., Zheng, X.-B., Luo, X.-J., & Mai, B.-X. (2011). Brominated flame retardants in the atmosphere of e-waste and rural sites in southern China: Seasonal variation, temperature dependence, and gas-particle partitioning. *Environmental Science and Technology, 45*(20), 8819–8825.

Tsydenova, O., & Bengtsson, M. (2011). Chemical hazards associated with treatment of waste electrical and electronic equipment. *Waste Management, 31*(1), 45–58.

Ullrich, S. M., Tanton, T. W., & Abdrashitova, S. A. (2001). Mercury in the aquatic environment: A review of factors affecting methylation. *Critical Reviews in Environmental Science and Technology, 31*(3), 241–293.

Uusküla, M., Järventaus, H., Hirvonen, A., Sorsa, M., & Norppa, H. (1995). Influence of GSTM1 genotype on sister chromatid exchange induction by styrene-7, 8-oxide and 1, 2-epoxy-3-butene in cultured human lymphocytes. *Carcinogenesis, 16*(4), 947–950.

Vahter, M., Åkesson, A., Lidén, C., Ceccatelli, S., & Berglund, M. (2007). Gender differences in the disposition and toxicity of metals. *Environmental Research, 104*(1), 85–95.

Van den Berg, M., Birnbaum, L. S., Denison, M., De Vito, M., Farland, W., Feeley, M., … Peterson, R. E. (2006). The 2005 World Health Organization reevaluation of human and mammalian toxic equivalency factors for dioxins and dioxin-like compounds. *Toxicological Sciences, 93*(2), 223–241.

Verity, M., Brown, W., Cheung, M., & Czer, G. (1977). Methyl mercury inhibition of synaptosome and brain slice protein synthesis: In vivo and in vitro studies. *Journal of Neurochemistry, 29*(4), 673–679.

Virolainen, S., Ibana, D., & Paatero, E. (2011). Recovery of indium from indium tin oxide by solvent extraction. *Hydrometallurgy, 107*(1), 56–61.

Waalkes, M. P. (2000). Cadmium carcinogenesis in review. *Journal of Inorganic Biochemistry, 79*(1), 241–244.

Waisberg, M., Joseph, P., Hale, B., & Beyersmann, D. (2003). Molecular and cellular mechanisms of cadmium carcinogenesis. *Toxicology, 192*(2), 95–117.

Wang, Y., Fang, J., Leonard, S. S., & Rao, K. M. K. (2004). Cadmium inhibits the electron transfer chain and induces reactive oxygen species. *Free Radical Biology and Medicine, 36*(11), 1434–1443.

Wang, Y., Luo, C., Li, J., Yin, H., Li, X., & Zhang, G. (2011). Characterization of PBDEs in soils and vegetations near an e-waste recycling site in South China. *Environmental Pollution, 159*(10), 2443–2448.

Wang, Y., Luo, C.-L., Li, J., Yin, H., Li, X.-D., & Zhang, G. (2011). Characterization and risk assessment of polychlorinated biphenyls in soils and vegetations near an electronic waste recycling site, South China. *Chemosphere, 85*(3), 344–350.

Wang, J., Ma, Y.-J., Chen, S.-J., Tian, M., Luo, X.-J., & Mai, B.-X. (2010). Brominated flame retardants in house dust from e-waste recycling and urban areas in South China: Implications on human exposure. *Environment International, 36*(6), 535–541.

Wang, L., Nagesha, D. K., Selvarasah, S., Dokmeci, M. R., & Carrier, R. L. (2008). Toxicity of CdSe nanoparticles in Caco-2 cell cultures. *Journal of Nanobiotechnology, 6*(1), 1.

Wang, Y., Tian, Z., Zhu, H., Cheng, Z., Kang, M., Luo, C., … Zhang, G. (2012). Polycyclic aromatic hydrocarbons (PAHs) in soils and vegetation near an e-waste recycling site in South China: Concentration, distribution, source, and risk assessment. *Science of the Total Environment, 439,* 187–193.

Wang, R., & Xu, Z. (2014). Recycling of non-metallic fractions from waste electrical and electronic equipment (WEEE): A review. *Waste Management, 34*(8), 1455–1469.

Webb, D., Sipes, I., & Carter, D. (1984). In vitro solubility and in vivo toxicity of gallium arsenide. *Toxicology and Applied Pharmacology, 76*(1), 96–104.

Webb, D., Wilson, S., & Carter, D. (1987). Pulmonary clearance and toxicity of respirable gallium arsenide particulates intratracheally instilled into rats. *The American Industrial Hygiene Association Journal, 48*(7), 660–667.

Wong, M., Wu, S., Deng, W., Yu, X., Luo, Q., Leung, A., … Wong, A. (2007). Export of toxic chemicals – a review of the case of uncontrolled electronic-waste recycling. *Environmental Pollution, 149*(2), 131–140.

Woods, J. S., Carver, G. T., & Fowler, B. A. (1979). Altered regulations of hepatic heme metabolism by indium chloride. *Toxicology and Applied Pharmacology, 49,* 455–461.

Wu, J.-P., Luo, X.-J., Zhang, Y., Luo, Y., Chen, S.-J., Mai, B.-X., & Yang, Z.-Y. (2008). Bioaccumulation of polybrominated diphenyl ethers (PBDEs) and polychlorinated biphenyls (PCBs) in wild aquatic species from an electronic waste (e-waste) recycling site in South China. *Environment International, 34*(8), 1109–1113.

Yamauchi, H., Takahashi, K., & Yamamura, Y. (1986). Metabolism and excretion of orally and intraperitoneally administered gallium arsenide in the hamster. *Toxicology, 40*(3), 237–246.

Yamauchi, H., Takahashi, K., Yamamura, Y., & Fowler, B. A. (1992). Metabolism of subcutaneous administered indium arsenide in the hamster. *Toxicology and Applied Pharmacology, 116*(1), 66–70.

Yan, C., Dadvand, A., Rosei, F., & Perepichka, D. F. (2010). Near-IR photoresponse in new up-converting CdSe/NaYF$_4$: Yb, Er nanoheterostructures. *Journal of the American Chemical Society, 132*(26), 8868–8869.

Yan, M., Zhang, Y., Xu, K., Fu, T., Qin, H., & Zheng, X. (2011). An in vitro study of vascular endothelial toxicity of CdTe quantum dots. *Toxicology, 282*(3), 94–103.

Yang, J., Retegan, T., & Ekberg, C. (2013). Indium recovery from discarded LCD panel glass by solvent extraction. *Hydrometallurgy, 137,* 68–77.

Yang, M., & Chitambar, C. R. (2008). Role of oxidative stress in the induction of metallothionein-2A and heme oxygenase-1 gene expression by the antineoplastic agent gallium nitrate in human lymphoma cells. *Free Radical Biology and Medicine, 45*(6), 763–772.

Yasutake, A., Nakano, A., & Hirayama, K. (1998). Induction by mercury compounds of brain metallothionein in rats: Hg^0 exposure induces long-lived brain metallothionein. *Archives of Toxicology, 72*(4), 187–191.

Zhang, Y., Chen, W., Zhang, J., Liu, J., Chen, G., & Pope, C. (2007). In vitro and in vivo toxicity of CdTe nanoparticles. *Journal of Nanoscience and Nanotechnology, 7*(2), 497–503.

Zhang, B.-Z., Guan, Y.-F., Li, S.-M., & Zeng, E. Y. (2009). Occurrence of polybrominated diphenyl ethers in air and precipitation of the Pearl River Delta, South China: Annual washout ratios and depositional rates. *Environmental Science and Technology, 43*(24), 9142–9147.

Zhang, Q., Ye, J., Chen, J., Xu, H., Wang, C., & Zhao, M. (2014). Risk assessment of polychlorinated biphenyls and heavy metals in soils of an abandoned e-waste site in China. *Environmental Pollution, 185*, 258–265.

Zhao, X.-R., Qin, Z.-F., Yang, Z.-Z., Zhao, Q., Zhao, Y.-X., Qin, X.-F., … Xu, X.-B. (2010). Dual body burdens of polychlorinated biphenyls and polybrominated diphenyl ethers among local residents in an e-waste recycling region in Southeast China. *Chemosphere, 78*(6), 659–666.

Zhao, G., Zhou, H., Wang, D., Zha, J., Xu, Y., Rao, K., … Wang, Z. (2009). PBBs, PBDEs, and PCBs in foods collected from e-waste disassembly sites and daily intake by local residents. *Science of the Total Environment, 407*(8), 2565–2575.

Zheng, J., Chen, K.-H., Yan, X., Chen, S.-J., Hu, G.-C., Peng, X.-W., … Yang, Z.-Y. (2013). Heavy metals in food, house dust, and water from an e-waste recycling area in South China and the potential risk to human health. *Ecotoxicology and Environmental Safety, 96*, 205–212.

Zhong, H., Zhou, Y., Yang, Y., Yang, C., & Li, Y. (2007). Synthesis of type II CdTe-CdSe nanocrystal heterostructured multiple-branched rods and their photovoltaic applications. *The Journal of Physical Chemistry C, 111*(17), 6538–6543.

CHAPTER 4

Populations at Special Risk

1. IN UTERO EXPOSURE TO E-WASTE CHEMICALS

The placenta is an essential organ for delivering nutrients to the developing fetus (Cross, 2005). A number of the e-waste chemicals discussed in previous chapters are capable of crossing the placenta (Barr, Bishop, & Needham, 2007; Chen, Dietrich, Huo, & Ho, 2011; Needham et al., 2010), and hence in utero exposure of the developing fetus is a major concern. These chemicals may alter fetal development by altering cellular imprinting at the embryonic stage (Feil & Fraga, 2012; McLachlan, 2001; Murphy & Jirtle, 2000; Thompson et al., 2001), organ development during organogenesis (Bigsby et al., 1999; Daston et al., 1991; Rutledge, 1997), or fetal growth at later in utero stages of development (Markey, Luque, de Toro, Sonnenschein, & Soto, 2001; Sharara, Seifer, & Flaws, 1998). In China, marked adverse birth outcomes have been reported in an e-waste recycling area (Xu et al., 2012). The impact of impaired infants from birth on the resources of developing countries is not to be taken lightly.

2. CHILDREN

In many developing countries, children are living in villages involved in the recycling of e-waste materials (Huo et al., 2007; Wang et al., 2010; Zheng et al., 2008). E-waste recycling practices in developing countries may include breaking and shredding of components by children using hand tools (Brigden, Labunska, Santillo, & Johnston, 2008) and tending open-pit fires to remove PVC insulation from copper wiring and use of mercury to recover gold from circuit boards (Brigden et al., 2008; Caravanos, Clark, Fuller, & Lambertson, 2011; Ha et al., 2009). In either case, chemical exposures will occur in a growing and developing population and may result in further health difficulties later in life (Damstra, 2002; Gluckman, Hanson, Cooper, & Thornburg, 2008; Landrigan et al., 2005; Li et al., 2008). Further development of biomarkers linked to chemical biomonitoring studies to accurately assess putative relationships between early chemical exposures and later health outcomes (Boekelheide et al., 2012; Gore, Walker, Zama, Armenti, & Uzumcu, 2011; Schoeters et al.,

55

2011; Schug, Janesick, Blumberg, & Heindel, 2011) is urgently needed to provide mode of action risk assessment information to address this important public health issue.

3. ADULTS OF CHILDBEARING AGE

The ability to conceive and bear children is a major issue for young adults, and chemical-induced alterations in reproductive function are major public health issues in many countries (Xu et al., 2012). A number of e-waste chemicals are well-known endocrine disruptors that may interfere with the normal functioning of reproductive systems (Craig, Wang, & Flaws, 2011; Pflieger-Bruss, Schuppe, & Schill, 2004). Further studies that examine the biological interactions of endocrine-disrupting chemicals with human populations during their peak reproductive years are needed to understand how such interactions could produce adverse health outcomes in otherwise healthy adults and/or their offspring during this critical period of life.

4. ELDERLY

The elderly are another population at special risk due to reduced capacity to deal with chemical insults at this stage of life (Geller & Zenick, 2005; Ginsberg, Hattis, Russ, & Sonawane, 2005; Rikans, 1989; Rikans & Hornbrook, 1997). The impact of chemical exposures on geriatric populations is important in terms of producing adverse health outcomes but also exacerbating the effects of other adverse health risks such as infectious diseases (Luebke, Parks, & Luster, 2004). A number of e-waste chemicals such as cadmium (Fowler, 2013) are known to more severely affect elderly populations ostensibly due to a combination of factors such as decreased biological reserve capacity leading to a diminished ability to respond to chemical insults via synthesis of cadmium-binding proteins such as metallothionein. Dietary deficiencies in the intake of essential elements and proteins may also play an interactive role. This scenario could be of particular importance among persons engaged in e-waste recycling in developing countries.

5. GENETIC INHERITANCE

It is increasingly clear that genetic inheritance plays a major role in mediating susceptibility to chemical toxicity (Božina, Bradamante, & Lovrić, 2009) for a number of chemicals such as lead found in e-waste (Chen et al., 2010; Frazzoli, Orisakwe, Dragone, & Mantovani, 2010; Scinicariello, Yesupriya, Chang, & Fowler, 2010). This may occur as a result of altered chemical metabolism (Chen et al., 2010) or regulation of mechanisms of toxicity (Scinicariello et al., 2010). The roles of genetic inheritance in mediating susceptibility in toxic chemicals have been appreciated for many years (Miller, Mohrenweiser, & Bell, 2001; Thier et al., 2003), and it is clear that wide differences exist across general populations in sensitivity to chemical agents. On the other hand, it is worth noting that e-waste recycling in developing countries may be conducted in small localized communities and among extended families who are genetically related to each other. This means that such population groups may be a special risk to toxicity from e-waste chemicals, and risk assessment studies should consider incorporating an examination of such groups.

6. PERSONS OF POOR NUTRITIONAL STATUS

In many developing countries, access to food and the ability to pay for it are the major problems (Rosegrant & Cline, 2003; Rosegrant, Ringler, & Zhu, 2009). Poverty is a major driver for the localization of e-waste recycling in developing countries since it may provide a source of much needed funds. Poor nutritional status has been shown to play a role in mediating the effects of many of the chemicals found in the e-waste stream (Hennig et al., 2007; Mahaffey & Vanderveen, 1979). Poor nutrition may compromise metabolic chemical defense systems by reducing access to amino acids needed for their synthesis (Yang, Brady, & Hong, 1992). The combination of e-waste chemical exposures and decreased metabolic capacity creates a situation where individuals in this situation would have likely greater susceptibility to adverse health outcomes.

7. SUBSISTENCE FARMERS/HUNTERS AND FISHERS/ LOW SOCIOECONOMIC STATUS

In many developing countries, persons of low socioeconomic status must engage in subsistence farming and supplement their diets with wild fish and game. The localization of e-waste recycling facilities and open burn pits near villages where subsistence farming (Qu et al., 2007) and hunting/fishing occur increase the likelihood of human exposures to e-waste chemicals from such situations as discussed in previous chapters.

8. CONTAMINATION OF LOCAL FOOD SUPPLIES AND HOUSE DUST

As discussed in previous chapters, the colocalization of e-waste recycling facilities has been documented to result in chemical contamination of local livestock and produce consumed as food. Much of this contamination occurs as result of chemicals in particulate matter released during incineration of e-waste.

These particles also become components of house dust (Leung, Duzgoren-Aydin, Cheung, & Wong, 2008; Sjödin et al., 2008; Tue et al., 2010; Wang et al., 2010). Persons living in such circumstances and engaged in e-waste recycling practices are hence at increased risk of chemical toxicity due to occupational, environmental, and dietary exposures (Bi et al., 2007; Grant et al., 2013; Ma et al., 2008).

References

Barr, D. B., Bishop, A., & Needham, L. L. (2007). Concentrations of xenobiotic chemicals in the maternal-fetal unit. *Reproductive Toxicology, 23*(3), 260–266.

Bi, X., Thomas, G. O., Jones, K. C., Qu, W., Sheng, G., Martin, F. L., & Fu, J. (2007). Exposure of electronics dismantling workers to polybrominated diphenyl ethers, polychlorinated biphenyls, and organochlorine pesticides in South China. *Environmental Science and Technology, 41*(16), 5647–5653.

Bigsby, R., Chapin, R. E., Daston, G. P., Davis, B. J., Gorski, J., Gray, L. E., … Vom Saal, F. S. (1999). Evaluating the effects of endocrine disruptors on endocrine function during development. *Environmental Health Perspectives, 107*(Suppl. 4), 613.

Boekelheide, K., Blumberg, B., Chapin, R. E., Cote, I., Graziano, J. H., Janesick, A., … Rogers, J. M. (2012). Predicting later-life outcomes of early-life exposures. *Environmental Health Perspectives, 120*(10), 1353.

Božina, N., Bradamante, V., & Lovrić, M. (2009). Genetic polymorphism of metabolic enzymes P450 (CYP) as a susceptibility factor for drug response, toxicity, and cancer risk. *Archives of Industrial Hygiene and Toxicology, 60*(2), 217–242.

Brigden, K., Labunska, I., Santillo, D., & Johnston, P. (2008). *Chemical contamination at e-waste recycling and disposal sites in Accra and Korforidua.* Ghana, Amsterdam: Greenpeace.

Caravanos, J., Clark, E., Fuller, R., & Lambertson, C. (2011). Assessing worker and environmental chemical exposure risks at an e-waste recycling and disposal site in Accra, Ghana. *Journal of Health and Pollution, 1*(1), 16–25.

Chen, A., Dietrich, K. N., Huo, X., & Ho, S.-M. (2011). Developmental neurotoxicants in e-waste: An emerging health concern. *Environmental Health Perspectives, 119*(4), 431.

Chen, L., Guo, H., Yuan, J., He, M., Chen, D., Shi, J., … Chen, X. (2010). Polymorphisms of GSTT1 and GSTM1 and increased micronucleus frequencies in peripheral blood lymphocytes in residents at an e-waste dismantling site in China. *Journal of Environmental Science and Health, Part A, 45*(4), 490–497.

Craig, Z. R., Wang, W., & Flaws, J. A. (2011). Endocrine-disrupting chemicals in ovarian function: Effects on steroidogenesis, metabolism and nuclear receptor signaling. *Reproduction, 142*(5), 633–646.

Cross, J. C. (2005). Placental function in development and disease. *Reproduction, Fertility and Development, 18*(2), 71–76.

Damstra, T. (2002). Potential effects of certain persistent organic pollutants and endocrine disrupting chemicals on the health of children. *Journal of Toxicology: Clinical Toxicology, 40*(4), 457–465.

Daston, G. P., Overmann, G. J., Taubeneck, M. W., Lehman-McKeeman, L. D., Rogers, J. M., & Keen, C. L. (1991). The role of metallothionein induction and altered zinc status in maternally mediated developmental toxicity: Comparison of the effects of urethane and styrene in rats. *Toxicology and Applied Pharmacology, 110*(3), 450–463.

Feil, R., & Fraga, M. F. (2012). Epigenetics and the environment: Emerging patterns and implications. *Nature Reviews Genetics, 13*(2), 97–109.

Fowler, B. A. (Ed.). (2013). *Computational toxicology: Applications for risk assessment* (pp. 258). Amsterdam: Elsevier Publishers.

Frazzoli, C., Orisakwe, O. E., Dragone, R., & Mantovani, A. (2010). Diagnostic health risk assessment of electronic waste on the general population in developing countries' scenarios. *Environmental Impact Assessment Review, 30*(6), 388–399.

Geller, A. M., & Zenick, H. (2005). Aging and the environment: A research framework. *Environmental Health Perspectives*, 1257–1262.

Ginsberg, G., Hattis, D., Russ, A., & Sonawane, B. (2005). Pharmacokinetic and pharmacodynamic factors that can affect sensitivity to neurotoxic sequelae in elderly individuals. *Environmental Health Perspectives*, 1243–1249.

Gluckman, P. D., Hanson, M. A., Cooper, C., & Thornburg, K. L. (2008). Effect of in utero and early-life conditions on adult health and disease. *New England Journal of Medicine, 359*(1), 61–73.

Gore, A. C., Walker, D. M., Zama, A. M., Armenti, A. E., & Uzumcu, M. (2011). Early life exposure to endocrine-disrupting chemicals causes lifelong molecular reprogramming of the hypothalamus and premature reproductive aging. *Molecular Endocrinology, 25*(12), 2157–2168.

Grant, K., Goldizen, F. C., Sly, P. D., Brune, M.-N., Neira, M., van den Berg, M., & Norman, R. E. (2013). Health consequences of exposure to e-waste: A systematic review. *The Lancet Global Health, 1*(6), e350–e361.

Ha, N. N., Agusa, T., Ramu, K., Tu, N. P. C., Murata, S., Bulbule, K. A., … Tanabe, S. (2009). Contamination by trace elements at e-waste recycling sites in Bangalore, India. *Chemosphere, 76*(1), 9–15.

Hennig, B., Ettinger, A. S., Jandacek, R. J., Koo, S., McClain, C., Seifried, H., … Suk, W. A. (2007). Using nutrition for intervention and prevention against environmental chemical toxicity and associated diseases. *Environmental Health Perspectives*, 493–495.

Huo, X., Peng, L., Xu, X., Zheng, L., Qiu, B., Qi, Z., … Piao, Z. (2007). Elevated blood lead levels of children in Guiyu, an electronic waste recycling town in China. *Environmental Health Perspectives*, 1113–1117.

Landrigan, P. J., Sonawane, B., Butler, R. N., Trasande, L., Callan, R., & Droller, D. (2005). Early environmental origins of neurodegenerative disease in later life. *Environmental Health Perspectives*, 1230–1233.

Leung, A. O., Duzgoren-Aydin, N. S., Cheung, K., & Wong, M. H. (2008). Heavy metals concentrations of surface dust from e-waste recycling and its human health implications in southeast China. *Environmental Science and Technology, 42*(7), 2674–2680.

Li, Y., Xu, X., Wu, K., Chen, G., Liu, J., Chen, S., … Huo, X. (2008). Monitoring of lead load and its effect on neonatal behavioral neurological assessment scores in Guiyu, an electronic waste recycling town in China. *Journal of Environmental Monitoring, 10*(10), 1233–1238.

Luebke, R. W., Parks, C., & Luster, M. I. (2004). Suppression of immune function and susceptibility to infections in humans: Association of immune function with clinical disease. *Journal of Immunotoxicology, 1*(1), 15–24.

Ma, J., Kannan, K., Cheng, J., Horii, Y., Wu, Q., & Wang, W. (2008). Concentrations, profiles, and estimated human exposures for polychlorinated dibenzo-p-dioxins and dibenzofurans from electronic waste recycling facilities and a chemical industrial complex in Eastern China. *Environmental Science and Technology, 42*(22), 8252–8259.

Mahaffey, K. R., & Vanderveen, J. E. (1979). Nutrient-toxicant interactions: Susceptible populations. *Environmental Health Perspectives, 29*, 81.

Markey, C. M., Luque, E. H., de Toro, M. M., Sonnenschein, C., & Soto, A. M. (2001). In utero exposure to bisphenol A alters the development and tissue organization of the mouse mammary gland. *Biology of Reproduction, 65*(4), 1215–1223.

McLachlan, J. A. (2001). Environmental signaling: What embryos and evolution teach us about endocrine disrupting chemicals. *Endocrine Reviews, 22*(3), 319–341.

Miller, M. C., Mohrenweiser, H. W., & Bell, D. A. (2001). Genetic variability in susceptibility and response to toxicants. *Toxicology Letters, 120*(1), 269–280.

Murphy, S. K., & Jirtle, R. L. (2000). Imprinted genes as potential genetic and epigenetic toxicologic targets. *Environmental Health Perspectives, 108*(Suppl. 1), 5.

Needham, L. L., Grandjean, P., Heinzow, B., Jørgensen, P. J., Nielsen, F., Patterson, D. G., Jr., … Weihe, P. (2010). Partition of environmental chemicals between maternal and fetal blood and tissues. *Environmental Science and Technology, 45*(3), 1121–1126.

Pflieger-Bruss, S., Schuppe, H. C., & Schill, W. B. (2004). The male reproductive system and its susceptibility to endocrine disrupting chemicals. *Andrologia, 36*(6), 337–345.

Qu, W., Bi, X., Sheng, G., Lu, S., Fu, J., Yuan, J., & Li, L. (2007). Exposure to polybrominated diphenyl ethers among workers at an electronic waste dismantling region in Guangdong, China. *Environment International, 33*(8), 1029–1034.

Rikans, L. E. (1989). Influence of aging on chemically induced hepato-toxicity: Role of age-related changes in metabolism. *Drug Metabolism Reviews, 20*(1), 87–110.

Rikans, L. E., & Hornbrook, K. R. (1997). Age-related susceptibility to hepatotoxicants. *Environmental Toxicology and Pharmacology, 4*(3), 339–344.

Rosegrant, M. W., & Cline, S. A. (2003). Global food security: Challenges and policies. *Science, 302*(5652), 1917–1919.

Rosegrant, M. W., Ringler, C., & Zhu, T. (2009). Water for agriculture: Maintaining food security under growing scarcity. *Annual Review of Environment and Resources, 34*(1), 205.

Rutledge, J. C. (1997). Developmental toxicity induced during early stages of mammalian embryogenesis. *Mutation Research/Fundamental and Molecular Mechanisms of Mutagenesis, 396*(1), 113–127.

Schoeters, G. E., Den Hond, E., Koppen, G., Smolders, R., Bloemen, K., De Boever, P., & Govarts, E. (2011). Biomonitoring and biomarkers to unravel the risks from prenatal environmental exposures for later health outcomes. *The American Journal of Clinical Nutrition, 94*(Suppl. 6), 1964S–1969S.

Schug, T. T., Janesick, A., Blumberg, B., & Heindel, J. J. (2011). Endocrine disrupting chemicals and disease susceptibility. *The Journal of Steroid Biochemistry and Molecular Biology, 127*(3), 204–215.

Scinicariello, F., Yesupriya, A., Chang, M.-H., & Fowler, B. A. (2010). Modification by ALAD of the association between blood lead and blood pressure in the US population: Results from the Third National Health and Nutrition Examination Survey. *Environmental Health Perspectives, 118*(2), 259.

Sharara, F. I., Seifer, D. B., & Flaws, J. A. (1998). Environmental toxicants and female reproduction. *Fertility and Sterility, 70*(4), 613–622.

Sjödin, A., Päpke, O., McGahee, E., Focant, J.-F., Jones, R. S., Pless-Mulloli, T., … Patterson, D. G., Jr. (2008). Concentration of polybrominated diphenyl ethers (PBDEs) in household dust from various countries. *Chemosphere, 73*(1), S131–S136.

Thier, R., Brüning, T., Roos, P. H., Rihs, H.-P., Golka, K., Ko, Y., & Bolt, H. M. (2003). Markers of genetic susceptibility in human environmental hygiene and toxicology: The role of selected CYP, NAT and GST genes. *International Journal of Hygiene and Environmental Health, 206*(3), 149–171.

Thompson, S. L., Konfortova, G., Gregory, R. I., Reik, W., Dean, W., & Feil, R. (2001). Environmental effects on genomic imprinting in mammals. *Toxicology Letters, 120*(1), 143–150.

Tue, N. M., Suzuki, G., Takahashi, S., Isobe, T., Trang, P. T., Viet, P. H., & Tanabe, S. (2010). Evaluation of dioxin-like activities in settled house dust from Vietnamese e-waste recycling sites: Relevance of polychlorinated/brominated dibenzo-p-dioxin/furans and dioxin-like PCBs. *Environmental Science and Technology*, *44*(23), 9195–9200.

Wang, J., Ma, Y.-J., Chen, S.-J., Tian, M., Luo, X.-J., & Mai, B.-X. (2010). Brominated flame retardants in house dust from e-waste recycling and urban areas in South China: Implications on human exposure. *Environment International*, *36*(6), 535–541.

Xu, X., Yang, H., Chen, A., Zhou, Y., Wu, K., Liu, J., ... Huo, X. (2012). Birth outcomes related to informal e-waste recycling in Guiyu, China. *Reproductive Toxicology*, *33*(1), 94–98.

Yang, C. S., Brady, J. F., & Hong, J.-Y. (1992). Dietary effects on cytochromes P450, xenobiotic metabolism, and toxicity. *The FASEB Journal*, *6*(2), 737–744.

Zheng, L., Wu, K., Li, Y., Qi, Z., Han, D., Zhang, B., ... Huo, X. (2008). Blood lead and cadmium levels and relevant factors among children from an e-waste recycling town in China. *Environmental Research*, *108*(1), 15–20.

CHAPTER 5

Risk Assessment/Risk Communication Approaches for E-Waste Sites

1. INDIVIDUAL CHEMICAL APPROACHES

Historically, risk assessments for chemical toxicity have been performed on a one chemical at a time approach (Fryer, Collins, Ferrier, Colvile, & Nieuwenhuijsen, 2006), which is useful if only one chemical dominates the known exposure pattern. Many of the chemicals which fall into this category are high production volume chemicals (Heidorn, Hansen, & Nørager, 1996; Sanderson et al., 2009), which have been studied for a number of decades and whose toxic properties are well known. This is, however, not the usual situation and especially so for e-waste chemical exposures that are mixtures of both organic and metallic chemical species (Haddad, Béliveau, Tardif, & Krishnan, 2001; Teuschler, 2007). This situation is being further complicated by the fabrication of nanoparticle formulations of these agents (Rushton et al., 2010) and the diverse nature of the workforce engaged in e-waste recycling that will vary on the basis of age, gender, nutritional status, socio-economic status (SES), and genetic inheritance (Bellinger, 2008; Ginsberg, Slikker, Bruckner, & Sonawane, 2004; Krewski et al., 2014; Pan et al., 2006; Perera et al., 2002). It should be clear from this discussion that a "one size fits all" risk assessment approach is not appropriate for dealing e-waste chemical exposures. Nonetheless, past experience with analogous Superfund site situations (Labieniec, Dzombak, & Siegrist, 1997; Linkov et al., 2006) does provide some useful guidance in terms of how to approach e-waste recycling sites involving new classes of chemical agents with regard to appropriate questions to be considered and needed analytical/technical information to inform a rigorous site risk assessment.

2. MIXTURE APPROACHES

As noted above, chemicals in the e-waste stream most frequently occur as mixtures (Backhaus & Faust, 2012; El-Masri, Thomas, Benjamin, & Yang, 1995; Teuschler, 2007). ATSDR has worked to develop methods for dealing with chemical mixtures using a binary weight of evidence approach

(Adeodato, Salazar, Gallindo, Sá, & Souza, 2014; Pohl, Fay, & Risher, 2005; Wilbur, Hansen, Pohl, Colman, & McClure, 2004), since chemical mixtures are also the most frequent situation commonly encountered in Superfund sites (Johnson & DeRosa, 1995; Monosson, 2005). The tools and approaches developed over the past decades for conducting risk assessments on Superfund sites are a useful starting place for conducting risk assessments on e-waste recycling sites. It should be noted that Superfund site approaches are somewhat limited in terms of including modern measures of toxicity such as omic biomarkers that more precisely define sensitive human subpopulations on the basis of age, gender, nutritional status, SES, nanomaterials, and genetic inheritance as noted above. Clearly there is more work to be done with regard to fine-tuning risk assessment strategies for e-waste situations as rapidly evolving environmental situations involving new types of chemical agents and sensitive populations living in developing countries (Jeyaratnam, 1990; Ortiz et al., 2002; Smith, Samet, Romieu, & Bruce, 2000). More recently, mixture risk assessments have incorporated molecular biomarker endpoints such as those outlined in the EPA NEXGEN Program (see Cote et al., 2016; Fowler, 2016) and Computational Toxicology modeling approaches (Fowler, 2016) for dealing with the large quantities of data generated by such multidisciplinary approaches.

Specifically, over the past 40 years, great interest has developed in the field of molecular biomarkers (see Fowler, 2016) for detection of early manifestations of cellular toxicity prior to the onset of overt organ system toxicity or cancer (Bonventre, Vaidya, Schmouder, Feig, & Dieterle, 2010; Cho, 2010; Hartwell, Mankoff, Paulovich, Ramsey, & Swisher, 2006; Joshi, Kaur, & Kaur, 2016; Nicholson, Connelly, Lindon, & Holmes, 2002; Reo, 2002). The "omic" biomarkers (genomic, proteomic, and metabolomic) are currently the main classes of biomarkers in play. Among the advantages of these early cellular responses to chemical-induced cell injury are that they are relatively inexpensive, rapid, and amenable to computer-managed automation and complimentary with automated chemical analytical equipment so that correlations may be made that link chemical dosimetry with a specific biological response (Fowler, 2009; MacGregor, 2003; Paules, 2003; Waters, 2003). These direct correlations open the door to mode of action (MOA)–based risk assessments (Borgert, Quill, McCarty, & Mason, 2004; Edwards & Preston, 2008). The USEPA has taken a strong leadership role in developing these approaches through its NEXGEN risk assessment initiative (Cote et al., 2012; Krewski et al., 2014; Zeise et al., 2013). It is clear that the application of new molecular tools may provide much needed insights for increasing the precision of risk assessments at e-waste sites in developing countries (Fowler, 2013).

3. AGE, NUTRITIONAL, GENDER, SOCIOECONOMIC STATUS, AND GENETIC SUSCEPTIBILITY FACTORS

As discussed elsewhere in this book, there are number of uncertainty factors (UFs) that must be considered in any chemical risk assessment, and this is particularly true for e-waste chemicals. Biological factors such as age, nutritional status, gender, and genetic susceptibility must also be considered in performing an accurate risk assessment evaluation for populations exposed to e-waste chemicals at an early life stage and but not experiencing adverse health effects such as obesity and diabetes (Grun & Blumberg, 2006; Reilly & Kelly, 2011; Thayer, Heindel, Bucher, & Gallo, 2012) until later in adulthood (Boekelheide et al., 2012; Fenton, 2006). In addition, the issue of low SES itself (Conroy, Sandel, & Zuckerman, 2010) should be considered in the calculation risk since poverty in developing countries is a major inducement for persons to engage in e-waste recycling under primitive conditions, which would only add to the health risks placed on these populations for development of chronic diseases or cancer later in life. There are studies (Landrigan et al., 1999) that have shown adverse health effects in populations who experienced chemical exposures during an impoverished childhood but subsequently moved out of poverty but who suffered adverse health effects later in life as a consequence of childhood experiences. The combined impact of poverty with unregulated chronic chemical exposures on long-term health outcomes should hence not be overlooked (Chi, Streicher-Porte, Wang, & Reuter, 2011; Nnorom & Osibanjo, 2008; Sinha-Khetriwal, Kraeuchi, & Schwaninger, 2005). The legacy of early childhood chemical exposures in combination with low SES during childhood may exert profound public health effects later in adult life. This will only be further exacerbated as the population of the world continues to age due to the reduction of infectious diseases via vaccination programs and use of antibiotics.

4. PERCEPTIONS OF RISK AT TOXIC WASTE SITES IN RELATION TO ECONOMIC AND FOOD CONCERNS: THE ROLE OF RISK COMMUNICATION

It is commonly the case that persons living near toxic waste sites, including e-waste sites, may be of low SES and have limited choices in terms of their life situation. This circumstance may also influence the perceptions of resident populations (Slovic, 1999) regarding chemical risks from their local environment if they are in need to earn a living in that locale and are subsistence hunters or fishers (Burger & Gochfeld, 1991) and are consuming locally grown crops (Cambra, Martínez, Urzelai, & Alonso, 1999). Low SES populations may hence have different responses to risk assessment information, and hence care must be taken to provide risk communication information in both a clear manner but also in terms that are culturally and linguistically acceptable so that no offense is taken and the population of concern is amenable to working with public health personnel to help resolve the situation in an expeditious manner.

5. COMPUTATIONAL TOXICOLOGY APPROACHES

Computational toxicology approaches to risk assessment are taken up in the first volume of this book series (Fowler, 2013). The great value of these tools rests with their speed, relatively low cost, and ability to integrate exposure and response data into useful information (Collins, Gray, & Bucher, 2008) for making health and regulatory decisions (Kavlock & Dix, 2010). In addition, computational modeling tools have proven of great value in data mining of existing health databases such as the CDC NHANES program to delineate subpopulations at special risk (females) in the general population following environmental exposures to cadmium (Ruiz, Mumtaz, Osterloh, Fisher, & Fowler, 2010). In addition, computational modeling techniques such as the application of Monte Carlo models have been used to help define predicted chemical exposures at Superfund sites (Simon, 1999; Smith, 1994). This type of information is very valuable in helping to focus more costly laboratory studies on sensitive subpopulations and expediting MOA risk assessment approaches (Boobis et al., 2006; Edwards & Preston, 2008; Sonich-Mullin et al., 2001) and expediting calculations that incorporate data on the uncertainty factors noted above so that these considerations may be backed up with actual data versus administrative procedures that have little basis in science.

In addition, the advent of System Biology approaches (Edwards & Preston, 2008; Ekins, Nikolsky, & Nikolskaya, 2005) to evaluating diseases has permitted new insights into evaluating the possible roles of chemicals in mediating diseases such as cancer (Hood, Heath, Phelps, & Lin, 2004; Kreeger & Lauffenburger, 2010) and metabolic disorders such as obesity and type II diabetes (Ruiz, Perlina, Mumtaz, & Fowler, 2016). The application of such computational toxicology tools has permitted the development of "knowledge maps" (Fig. 5.1), which outline possible relationships between specific chemical exposures and disruption of essential endocrine-mediated regulatory pathways. The value of such information is to guide and expedite laboratory-based research studies and reduce attendant research and development costs.

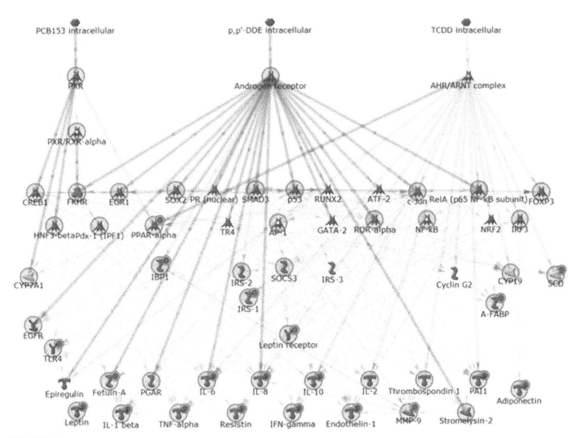

FIGURE 5.1

Proposed global network for potential converging genes associated with diabetes/insulin resistance, obesity, or metabolic syndrome X, and the three POPs. *Thick lines* highlight the closest interactions. *Large gray circles* represent union genes. *Small red circles* (small black circles in print version) indicate the intersection genes for the three diseases. Symbols defined by MetaCore at http://lsresearch.thomsonreuters.com/static/uploads/files/2014-05/MetaCoreQuickReferenceGuide.pdf. *Blue* (𝟐), receptors and adaptor proteins; *Green* (𝐓), cytokines and lipoproteins; *green arrows* (gray arrows in print version), activating interactions; *POPs*, hexagons; *Red* (𝖠), transcription factors; *red arrows* (black arrows in print version), inhibiting interactions; *Yellow* (◀), catalytic factors. *From Ruiz, P., Perlina, A., Mumtaz, M., & Fowler, B. A. (2016). A systems biology approach reveals converging molecular mechanisms that link different POPs to common metabolic diseases.* Environmental Health Perspectives, 124 *(7), 134–141.*

References

Adeodato, P. J., Salazar, D. S., Gallindo, L. S., Sá, Á. G., & Souza, S. M. (2014). Continuous variables segmentation and reordering for optimal performance on binary classification tasks. In *Paper presented at the 2014 International Joint Conference on Neural Networks (IJCNN)*.

Backhaus, T., & Faust, M. (2012). Predictive environmental risk assessment of chemical mixtures: A conceptual framework. *Environmental Science and Technology, 46*(5), 2564–2573.

Bellinger, D. C. (2008). Lead neurotoxicity and socioeconomic status: Conceptual and analytical issues. *Neurotoxicology, 29*(5), 828–832.

Boekelheide, K., Blumberg, B., Chapin, R. E., Cote, I., Graziano, J. H., Janesick, A., … Rogers, J. M. (2012). Predicting later-life outcomes of early-life exposures. *Environmental Health Perspectives, 120*(10), 1353.

Bonventre, J. V., Vaidya, V. S., Schmouder, R., Feig, P., & Dieterle, F. (2010). Next-generation bio-markers for detecting kidney toxicity. *Nature Biotechnology, 28*(5), 436.

Boobis, A. R., Cohen, S. M., Dellarco, V., McGregor, D., Meek, M., Vickers, C., … Farland, W. (2006). IPCS framework for analyzing the relevance of a cancer mode of action for humans. *Critical Reviews in Toxicology, 36*(10), 781–792.

Borgert, C. J., Quill, T. F., McCarty, L. S., & Mason, A. M. (2004). Can mode of action predict mix-ture toxicity for risk assessment? *Toxicology and Applied Pharmacology, 201*(2), 85–96.

Burger, J., & Gochfeld, M. (1991). Fishing a superfund site: Dissonance and risk perception of envi-ronmental hazards by fishermen in Puerto Rico. *Risk Analysis, 11*(2), 269–277.

Cambra, K., Martínez, T., Urzelai, A., & Alonso, E. (1999). Risk analysis of a farm area near a lead- and cadmium-contaminated industrial site. *Journal of Soil Contamination, 8*(5), 527–540.

Chi, X., Streicher-Porte, M., Wang, M. Y., & Reuter, M. A. (2011). Informal electronic waste recy-cling: A sector review with special focus on China. *Waste Management, 31*(4), 731–742.

Cho, W. C. (2010). MicroRNAs: Potential biomarkers for cancer diagnosis, prognosis and targets for therapy. *The International Journal of Biochemistry and Cell Biology, 42*(8), 1273–1281.

Collins, F. S., Gray, G. M., & Bucher, J. R. (2008). Transforming environmental health protection. *Science (New York), 319*(5865), 906.

Conroy, K., Sandel, M., & Zuckerman, B. (2010). Poverty grown up: How childhood socioeconomic status impacts adult health. *Journal of Developmental and Behavioral Pediatrics, 31*(2), 154–160.

Cote, I., Anastas, P. T., Birnbaum, L. S., Clark, R. M., Dix, D. J., Edwards, S. W., & Preuss, P. W. (2012). Advancing the next generation of health risk assessment. *Environmental Health Perspectives, 120*(11), 1499.

Cote, I., Andersen, M. E., Ankley, G. T., Barone, S., Birnbaum, L. S., Boekelheide,K., … DeWoskin, R. S. (2016). The next generation of risk assessment multiyear study-highlights of findings, applications to risk assessment, and future directions. *Environmental Health Perspectives, 124,* 1671–1682. http://dx.doi.org/10.1289/EHP233.

Edwards, S. W., & Preston, R. J. (2008). Systems biology and mode of action based risk assessment. *Toxicological Sciences, 106*(2), 312–318.

Ekins, S., Nikolsky, Y., & Nikolskaya, T. (2005). Techniques: Application of systems biology to absorption, distribution, metabolism, excretion and toxicity. *Trends in Pharmacological Sciences, 26*(4), 202–209.

El-Masri, H. A., Thomas, R. S., Benjamin, S. A., & Yang, R. S. (1995). Physiologically based phar-macokinetic/pharmacodynamic modeling of chemical mixtures and possible applications in risk assessment. *Toxicology, 105*(2), 275–282.

Fenton, S. E. (2006). Endocrine-disrupting compounds and mammary gland development: Early exposure and later life consequences. *Endocrinology, 147*(6), s18–s24.

Fowler, B. A. (2009). Monitoring of human populations for early markers of cadmium toxicity: A review. *Toxicology and Applied Pharmacology, 238*(3), 294–300.

Fowler, B. A. (2013). *Computational toxicology: Methods and applications for risk assessment.* Academic Press.

Fowler, B. A. (2016). *Molecular biological markers for toxicology and risk assessment* (pp. 153). Amsterdam: Elsevier Publishers.

Fryer, M., Collins, C. D., Ferrier, H., Colvile, R. N., & Nieuwenhuijsen, M. J. (2006). Human expo-sure modelling for chemical risk assessment: A review of current approaches and research and policy implications. *Environmental Science and Policy, 9*(3), 261–274.

Ginsberg, G., Slikker, W., Jr., Bruckner, J., & Sonawane, B. (2004). Incorporating children's toxicoki-netics into a risk framework. *Environmental Health Perspectives, 112*(2), 272.

Grun, F., & Blumberg, B. (2006). Environmental obesogens: Organotins and endocrine disruption via nuclear receptor signaling. *Endocrinology*, *147*(6), s50–s55.

Haddad, S., Béliveau, M., Tardif, R., & Krishnan, K. (2001). A PBPK modeling-based approach to account for interactions in the health risk assessment of chemical mixtures. *Toxicological Sciences*, *63*(1), 125–131.

Hartwell, L., Mankoff, D., Paulovich, A., Ramsey, S., & Swisher, E. (2006). Cancer biomarkers: A systems approach. *Nature Biotechnology*, *24*(8), 905–908.

Heidorn, C. J., Hansen, B. G., & Nørager, O. (1996). IUCLID: A database on chemical substances information as a tool for the EU-Risk-Assessment program. *Journal of Chemical Information and Computer Sciences*, *36*(5), 949–954.

Hood, L., Heath, J. R., Phelps, M. E., & Lin, B. (2004). Systems biology and new technologies enable predictive and preventative medicine. *Science*, *306*(5696), 640–643.

Jeyaratnam, J. (1990). Acute pesticide poisoning: A major global health problem. *World Health Statistics Quarterly*, *43*(3), 139–144.

Johnson, B. L., & DeRosa, C. T. (1995). Chemical mixtures released from hazardous waste sites: Implications for health risk assessment. *Toxicology*, *105*(2), 145–156.

Joshi, G., Kaur, R., & Kaur, H. (2016). Biomarkers in cancer.

Kavlock, R., & Dix, D. (2010). Computational toxicology as implemented by the US EPA: Providing high throughput decision support tools for screening and assessing chemical exposure, hazard and risk. *Journal of Toxicology and Environmental Health, Part B*, *13*(2–4), 197–217.

Kreeger, P. K., & Lauffenburger, D. A. (2010). Cancer systems biology: A network modeling perspective. *Carcinogenesis*, *31*(1), 2–8.

Krewski, D., Westphal, M., Andersen, M. E., Paoli, G. M., Chiu, W. A., Al-Zoughool, M., … Cote, I. (2014). A framework for the next generation of risk science. *Environmental Health Perspectives*, *122*(8), 796 (Online).

Labieniec, P. A., Dzombak, D. A., & Siegrist, R. L. (1997). Evaluation of uncertainty in a site-specific risk assessment. *Journal of Environmental Engineering*, *123*(3), 234–243.

Landrigan, P. J., Claudio, L., Markowitz, S. B., Berkowitz, G. S., Brenner, B. L., Romero, H., … Godbold, J. H. (1999). Pesticides and inner-city children: Exposures, risks, and prevention. *Environmental Health Perspectives*, *107*(Suppl. 3), 431.

Linkov, I., Satterstrom, F., Kiker, G., Batchelor, C., Bridges, T., & Ferguson, E. (2006). From comparative risk assessment to multi-criteria decision analysis and adaptive management: Recent developments and applications. *Environment International*, *32*(8), 1072–1093.

MacGregor, J. T. (2003). The future of regulatory toxicology: Impact of the biotechnology revolution. *Toxicological Sciences*, *75*(2), 236–248.

Monosson, E. (2005). Chemical mixtures: Considering the evolution of toxicology and chemical assessment. *Environmental Health Perspectives*, 383–390.

Nicholson, J. K., Connelly, J., Lindon, J. C., & Holmes, E. (2002). Metabonomics: A platform for studying drug toxicity and gene function. *Nature Reviews Drug Discovery*, *1*(2), 153–161.

Nnorom, I. C., & Osibanjo, O. (2008). Overview of electronic waste (e-waste) management practices and legislations, and their poor applications in the developing countries. *Resources, Conservation and Recycling*, *52*(6), 843–858.

Ortiz, D., Calderón, J., Batres, L., Carrizales, L., Mejía, J., Martínez, L., … Díaz-Barriga, F. (2002). Overview of human health and chemical mixtures: Problems facing developing countries. *Environmental Health Perspectives*, *110*(Suppl. 6), 901.

Pan, G., Hanaoka, T., Yoshimura, M., Zhang, S., Wang, P., Tsukino, H., … Takahashi, K. (2006). Decreased serum free testosterone in workers exposed to high levels of di-n-butyl phthalate (DBP) and di-2-ethylhexyl phthalate (DEHP): A cross-sectional study in China. *Environmental Health Perspectives*, 1643–1648.

Paules, R. (2003). Phenotypic anchoring: Linking cause and effect. *Environmental Health Perspectives, 111*(6), A338.

Perera, F. P., Illman, S. M., Kinney, P. L., Whyatt, R. M., Kelvin, E. A., Shepard, P., … Miller, R. L. (2002). The challenge of preventing environmentally related disease in young children: Community-based research in New York City. *Environmental Health Perspectives, 110*(2), 197.

Pohl, H., Fay, M., & Risher, J. (2005). Interaction profiles for simple mixtures: Mixtures with radioactive chemicals. *WIT Transactions on Ecology and the Environment, 85.*

Reilly, J. J., & Kelly, J. (2011). Long-term impact of overweight and obesity in childhood and adolescence on morbidity and premature mortality in adulthood: Systematic review. *International Journal of Obesity, 35*(7), 891–898.

Reo, N. V. (2002). NMR-based metabolomics. *Drug and Chemical Toxicology, 25*(4), 375–382.

Ruiz, P., Mumtaz, M., Osterloh, J., Fisher, J., & Fowler, B. A. (2010). Interpreting NHANES biomonitoring data, cadmium. *Toxicology Letters, 198*(1), 44–48.

Ruiz, P., Perlina, A., Mumtaz, M., & Fowler, B. A. (2016). A systems biology approach reveals converging molecular mechanisms that link different POPs to common metabolic diseases. *Environmental Health Perspectives, 124*(7), 134–141.

Rushton, E. K., Jiang, J., Leonard, S. S., Eberly, S., Castranova, V., Biswas, P., … Finkelstein, J. (2010). Concept of assessing nanoparticle hazards considering nanoparticle dosemetric and chemical/biological response metrics. *Journal of Toxicology and Environmental Health, Part A, 73*(5–6), 445–461.

Sanderson, H., Belanger, S. E., Fisk, P. R., Schäfers, C., Veenstra, G., Nielsen, A. M., … Stanton, K. (2009). An overview of hazard and risk assessment of the OECD high production volume chemical category—Long chain alcohols $[C_6–C_{22}]$ (LCOH). *Ecotoxicology and Environmental Safety, 72*(4), 973–979.

Simon, T. W. (1999). Two-dimensional Monte Carlo simulation and beyond: A comparison of several probabilistic risk assessment methods applied to a superfund site. *Human and Ecological Risk Assessment: An International Journal, 5*(4), 823–843.

Sinha-Khetriwal, D., Kraeuchi, P., & Schwaninger, M. (2005). A comparison of electronic waste recycling in Switzerland and in India. *Environmental Impact Assessment Review, 25*(5), 492–504.

Slovic, P. (1999). Trust, emotion, sex, politics, and science: Surveying the risk-assessment battlefield. *Risk Analysis, 19*(4), 689–701.

Smith, R. L. (1994). Use of Monte Carlo simulation for human exposure assessment at a superfund site. *Risk Analysis, 14*(4), 433–439.

Smith, K. R., Samet, J. M., Romieu, I., & Bruce, N. (2000). Indoor air pollution in developing countries and acute lower respiratory infections in children. *Thorax, 55*(6), 518–532.

Sonich-Mullin, C., Fielder, R., Wiltse, J., Baetcke, K., Dempsey, J., Fenner-Crisp, P., … Kroese, D. (2001). IPCS conceptual framework for evaluating a mode of action for chemical carcinogenesis. *Regulatory Toxicology and Pharmacology, 34*(2), 146–152.

Teuschler, L. K. (2007). Deciding which chemical mixtures risk assessment methods work best for what mixtures. *Toxicology and Applied Pharmacology, 223*(2), 139–147.

Thayer, K. A., Heindel, J. J., Bucher, J. R., & Gallo, M. A. (2012). Role of environmental chemicals in diabetes and obesity: A National Toxicology Program workshop review. *Environmental Health Perspectives, 120*(6), 779.

Waters, M. (2003). Systems toxicology and the chemical effects in biological systems (CEBS) knowledge base. *Environmental Health Perspectives, 111*(6), 811.

Wilbur, S. B., Hansen, H., Pohl, H., Colman, J., & McClure, P. (2004). Using the ATSDR guidance manual for the assessment of joint toxic action of chemical mixtures. *Environmental Toxicology and Pharmacology, 18*(3), 223–230.

Zeise, L., Bois, F. Y., Chiu, W. A., Hattis, D., Rusyn, I., & Guyton, K. Z. (2013). Addressing human variability in next-generation human health risk assessments of environmental chemicals. *Environmental Health Perspectives, 121*(1), 23 (Online).

CHAPTER 6

Translation of Risk Assessment Information Into Effective International Policies and Actions

1. COMMUNICATION OF SCIENTIFIC INFORMATION IN PRACTICAL TERMINOLOGY

For risk assessment information on e-waste to be of value to stakeholders, it must be translated into nontechnical or lay terms that can be understood by persons with more limited technical backgrounds (e.g., societal decision-makers /politicians). Mode of action (MOA) risk assessment is a key term that must be defined in lay terms for persons who need to understand the information but have limited technical backgrounds. One example may be government officials who need to make regulatory decisions but who are not versed in toxicology or risk assessment. Another important aspect is the need to translate risk assessment information into the native language of the country where e-waste recycling activities are being conducted. Many of these recycling activities are being conducted in rural areas under primitive conditions where formal education and use of English is limited so informational literature must be translated into the local language in lay terms in order to be effective. This is particularly important in developing countries where many of the persons engaged in e-waste recycling are low SES and may have little or no formal education or children who do not have access to proper formal educations (Dionisio et al., 2010; Zheng et al., 2008). For any e-waste risk assessment effort to be effective in protecting the public health, a serious communication effort must be made to reach such persons in a language they can understand. In practical terms, this will mean mounting a teaching effort starting with very basic concepts and explaining how the science is used to assess risk and why it is important for protecting the health of the e-waste recycling workers. One possible approach is to modify the ATSDR ToxFAQs information sheets which have been used effectively for a number of decades in relation to Superfund sites in the United States. These short highly effective communication documents have been translated into several languages and are aimed at the 8th grade reading level (see ATSDR, 2017). They provide a framework of essential information for the lay public and could be easily modified to incorporate new information on e-waste chemicals.

2. INFORMATION MAPPING TECHNOLOGY APPROACHES

One established approach for teaching lay persons about toxicology and risk assessment is a teaching technique named "information mapping technology" (Houghton, Pucar, & Knox, 1996). This tool is now used in school textbooks to communicate knowledge to students in an efficient manner. This approach has been adopted by ATSDR for a number of its communication products related to Superfund sites (Basak, Gute, Monteiro-Riviere, & Witzmann, 2010; Carr, 2003; Heitgerd, 2001) and specific toxic chemicals (ATSDR, 2007; Health & Services, 2000). The presented information is succinct and readily understandable at an 8th grade school level. The document poses a reasonable or likely question of concern related to the topic/chemical of concern and then provides a clear answer in plain English along with references that may be used if there are further questions or more information is required. ATSDR has incorporated this approach into a number of its documents such as ToxFAQs, Case Studies in Environmental Medicine, etc. Please see ATSDR (2017). A general teaching example of information mapping technology applied to a biomarker-based MOA risk assessment for arsenic in drinking water has been previously presented in another volume (Fowler, 2016) in this three-part book series. The point here is that information mapping technology is a powerful communication tool that could be effectively used in both developed and developing countries to provide needed practical risk assessment information on e-waste chemicals to persons and agencies needing such information to protect the public health. Access to this information will become only more important as the diaspora of discarded electronic devices to developing countries increases in volume and diversity of chemicals.

3. COLLABORATIONS AMONG INTERESTED INTERNATIONAL STAKEHOLDERS/GOVERNMENT AGENCIES/INDUSTRIAL GROUPS/NGOs

Given the global nature of the e-waste problem, it is essential that there be good and effective communication and collaboration among international stakeholders. These groups include government agencies, industrial groups, and NGOs with different interests in the e-waste problem. A concerted effort is needed to develop solid working relationships among these groups so they can coordinate with each other around common e-waste issues. This is not a small task since every organization, whether private or public, will have some intrinsic (Dahan, Doh, Oetzel, & Yaziji, 2010; Jonker & Nijhof, 2006; Stafford, Polonsky, & Hartman, 2000) bureaucratic and/or political structure that will need to be accommodated. An effective collaborative relationship between interested stakeholders requires time to develop trust to become effective and address public issues in developing countries with different cultural perspectives (Blaya, Fraser, & Holt, 2010; Ericson, Caravanos, Chatham-Stephens, Landrigan, & Fuller, 2013; Terazono et al., 2006). Nonetheless, there has been a solid movement for corporate social responsibility over the past 20 years in the area of environmental cooperation between corporate interests and NGOs and other stakeholders to look for solutions to the problems of common interest such as e-waste (Stafford et al., 2000). These a collaborations for the common good are heartening since they portend an awareness that all of us are living on a "spaceship," which, for the time being, is the sole and exclusive home for man and other species. If collectively we ruin our home through unwise stewardship, there may be potentially disastrous consequences on many levels with increasingly fewer options for remediation going forward in time. The e-waste problem is a clear and rapidly growing example of the need for improved cooperative approaches among stakeholders in various sectors to avert or at least attenuate a major environmental calamity. There is hence a pressing need to communicate to the scope and consequences of not addressing the e-waste problem in the near future to the general public and societal decision-makers in an effective manner not only using "plain English" but also translating the essential information into other languages to meet the needs of persons in other countries and cultures since the e-waste is a global problem that affects many countries.

4. INTERNATIONAL CONFERENCES AND DIPLOMATIC INTERACTIONS—BOTH FORMAL AND INFORMAL

Another potentially useful approach for communicating information about e-waste issues in both developed and developing countries is to sponsor conferences in the country of interest and invite the lay press. Such international conferences are most effectively conducted under the joint auspices of several agencies, groups, or NGOs with stakeholder interests. Such an approach would assure that information on the conference would reach all interested parties and if conducted in an evenhanded manner that all perspectives on the problem are heard.

The WHO, NATO, and UNEP are agencies that have a long track record of holding such meetings in relation to problems in developing countries (Achankeng, 2003; Terazono et al., 2006). In addition a number of US public such as the USEPA, NIEHS, and CDC/ATSDR have also sponsored conferences/meetings that have been focused on addressing emerging public health issues in developing countries (Nabel, Stevens, & Smith, 2009). E-waste public health concerns may be easily added to this list. A major value of such international face-to-face meetings is the opportunity for interpersonal networking and formation of direct lines of communication among the participants. Such interactions may lead to long-term working partnerships, which will be essential for addressing large complex problems on the scale of e-waste. This is essentially the functional basis of the "soft diplomacy" approach utilized by the Fulbright Program for increasing international understanding on complex issues (Nye, 2004, 2008). The global problem of e-waste is a classic example of where such a soft diplomatic approach could provide foundation for helping to resolve some of the core socioeconomic issues related to e-waste since there are many leaders in both developed and developing countries who are former Fulbright alumni and share a common understanding of the global community. It should also be noted here that the Fulbright Program is the most highly leveraged and effective international programs sponsored by the United States Government and has a remarkable record of international achievement for the reasons noted above. It is thus regrettable that this potent programmatic tool has suffered budget cuts in past decades, may be less able to help address major environmental problems such as e-waste due to diminished resources. The window of opportunity for dealing with major and growing international global problems such as e-waste, which has both socioeconomic and public health consequences, will not be open forever.

In addition, another important aspect of any conference is publication of the conference proceedings in highly visible vehicles. This may mean a high impact journal such as *Environmental Health Perspectives* or the websites for participating international agencies such as the WHO and UNEP or NGOs such as Greenpeace or scientific societies such as the Society of Toxicology. Social media and environmental blogs are additional sources of information to dissemination on major environmental issues such as e-waste. In any case, it is most important that information from such conferences be transparent and available to all interested parties in a timely manner and provide long-term record that may be updated over time to track progress in dealing with complex, large-scale, and long-term problems such as e-waste. Effective international communication is hence a key element for addressing e-waste and related issues.

References

Achankeng, E. (2003). Globalization, urbanization and municipal solid waste management in Africa. In *Paper presented at the proceedings of the African Studies Association of Australasia and the Pacific 26th Annual Conference.*

ATSDR, U. (2007). *Toxicological profile for lead* (Vol. 1. US Department of Health and Human Services, 582.

ATSDR (2017). Toxic substances portal. 4770 Buford Highway NE Atlanta, Georgia USA.

Basak, S. C., Gute, B. D., Monteiro-Riviere, N. A., & Witzmann, F. A. (2010). Characterization of toxicoproteomics maps for chemical mixtures using information theoretic approach. *Principles and Practice of Mixtures Toxicology*, 215–234.

Blaya, J. A., Fraser, H. S., & Holt, B. (2010). E-health technologies show promise in developing countries. *Health Affairs, 29*(2), 244–251.

Carr, T. (2003). Geographic information systems in the public sector. *Public Information Technology: Policy and Management Issues, 252.*

Dahan, N. M., Doh, J. P., Oetzel, J., & Yaziji, M. (2010). Corporate-NGO collaboration: Co-creating new business models for developing markets. *Long Range Planning, 43*(2), 326–342.

Dionisio, K. L., Arku, R. E., Hughes, A. F., Vallarino, J., Carmichael, H., Spengler, J. D., … Ezzati, M. (2010). Air pollution in Accra neighborhoods: Spatial, socioeconomic, and temporal patterns. *Environmental Science and Technology, 44*(7), 2270–2276.

Ericson, B., Caravanos, J., Chatham-Stephens, K., Landrigan, P., & Fuller, R. (2013). Approaches to systematic assessment of environmental exposures posed at hazardous waste sites in the developing world: The toxic sites identification program. *Environmental Monitoring and Assessment, 185*(2), 1755–1766.

Fowler, B. A. (2016). Molecular biological markers for toxicology and risk assessment (pp. 153). Amsterdam: Elsevier Publishers.

Health, U. D. o., & Services, H. (2000). *Toxicological profile for arsenic.*

Heitgerd, J. L. (2001). Using GIS and demographics to characterize communities at risk: A model for ATSDR. *Journal of Environmental Health, 64*(5), 21.

Houghton, J. W., Pucar, M., & Knox, C. (1996). Mapping information technology. *Futures, 28*(10), 903–917.

Jonker, J., & Nijhof, A. (2006). Looking through the eyes of others: Assessing mutual expectations and experiences in order to shape dialogue and collaboration between business and NGOs with respect to CSR. *Corporate Governance: An International Review, 14*(5), 456–466.

Nabel, E. G., Stevens, S., & Smith, R. (2009). Combating chronic disease in developing countries. *Lancet, 373*(9680), 2004–2006.

Nye, J. S., Jr. (2004). The decline of America's soft power-why Washington should worry. *Foreign Affairs, 83,* 16.

Nye, J. S. (2008). Public diplomacy and soft power. *The Annals of the American Academy of Political and Social Science, 616*(1), 94–109.

Stafford, E. R., Polonsky, M. J., & Hartman, C. L. (2000). Environmental NGO-business collaboration and strategic bridging: A case analysis of the Greenpeace-Foron alliance. *Business Strategy and the Environment, 9*(2), 122.

Terazono, A., Murakami, S., Abe, N., Inanc, B., Moriguchi, Y., Sakai, S.-I., … Williams, E. (2006). Current status and research on E-waste issues in Asia. *Journal of Material Cycles and Waste Management, 8*(1), 1–12.

Zheng, L., Wu, K., Li, Y., Qi, Z., Han, D., Zhang, B., … Huo, X. (2008). Blood lead and cadmium levels and relevant factors among children from an e-waste recycling town in China. *Environmental Research, 108*(1), 15–20.

CHAPTER 7

Current E-Waste Data Gaps and Future Research Directions

1. CURRENT GAPS IN THE E-WASTE DATABASE

1.1 General Introduction

There is a general appreciation that large quantities for discarded electronic devices are entering the e-waste stream and are being shipped to specific areas of a number of developing countries where they are being processed by unskilled labor, sometimes including children. Beyond this general picture based on estimates, there are only limited actual data on quantities and types of electronic devices being shipped for recycling and demographic information on workforce engaged in these unregulated recycling activities. Clearly, more specific information is needed to provide a basis for addressing the emerging environmental and public health issues linked to e-waste recycling in developing countries. This informal e-waste recycling situation has arisen by a confluence of socioeconomic factors, and it will take major international resources and cooperation to make progress on dealing with the global e-waste problem. It is important to begin the process now since the e-waste problem is growing larger and the types of materials entering into the e-waste steam are also changing (e.g., nanomaterials), thus further complicating public health risk assessments.

The following represent a short list of current data gaps and suggested areas of needed research that should help focus resources on productive avenues for addressing public health issues related to e-waste recycling in developing countries.

1.2 Paucity of Data on Quantities of Materials Entering the E-Waste Stream

Presently, there are few hard data on the quantities of discarded electronic devices entering the e-waste stream. Currently, only gross estimates of the types and quantities of such devices entering into informal recycling processes are available. A more rigorous accounting of the flow of such devices from developed to developing countries is needed. Such an accounting will be particularly important as new generations of devices with new materials

77

(e.g., nanomaterials) are introduced into commerce and become, in turn, obsolete. The global electronics industry is a rapidly evolving field of technology, and it can be expected that the chemicals incorporated into new devices, which will, in turn, become obsolete, will also change the character and magnitude of the associated chemical risks of new devices. Lack of data on changes in the types of materials entering the e-waste stream further complicates any the predictive capability of any risk assessments for e-waste.

1.3 A Lack of Hard Data on New Chemicals Entering the E-Waste Stream

There are a number of emerging toxicology and risk assessment issues associated with both the introduction and the evermore rapid turnover of chemicals in the e-waste stream. One emerging example may be liquid crystal alloys of gallium and indium, which permit soft "elastic" semiconductor applications such as solar energy conversion systems (Afzaal & O'Brien, 2006; Green, Emery, Hishikawa, Warta, & Dunlop, 2015; Lipomi & Bao, 2011). Such applications will allow for new devices, which presently do not exist but which can be expected to have a finite lifetime and ultimately need to be recycled.

1.4 Lack of Hard Information on Demographics of Persons Engaged in Recycling E-Waste

The toxicology issues associated with these rapidly evolving devices will provide new chemical challenges in relation to e-waste as they enter the e-waste stream with evermore rapid turnover. At present, it is known that much of the work of e-waste recycling is being conducted in developing countries by persons of low socioeconomic status including children. Actual hard data on the numbers, gender, and ages of these population groups are not available except as anecdotal observations. Clearly, hard survey data are needed to understand the magnitude of the problem and formulate wise policies regarding the use of child labor in the recycling of e-waste.

1.5 Limited Information on Current Public Health Resources in Developing Countries Related to E-Waste Recycling

There is also a limited amount of information on the public health resources available in developing countries to gather needed information on the scope of e-waste recycling and populations involved in these activities in developing countries. This will likely vary greatly between countries, but such information is needed to encourage the political leadership to provide resources in these countries and engage the international agencies to provide training and support to help address identified public health problem areas associated with e-waste.

2. FUTURE RESEARCH DIRECTIONS

2.1 Studies of Changes in Types and Quantities of E-Waste Arriving in Developing Countries

As noted above, the e-waste stream can be expected to change over time as new materials and chemicals are introduced into evolving new electronic devices (Kang & Schoenung, 2005). This will mean that the chemicals in the e-waste stream both in qualitative and quantitative terms thus present future challenges to toxicologists, risk assessors, and public health professionals in dealing with potential adverse toxicities or chemical-induced diseases resulting from exposures to new types of chemical agents (Kahhat et al., 2008; Plambeck & Wang, 2009; Streicher-Porte et al., 2005). An example of new chemicals that may be of concern is the nanomaterials being incorporated into the next generations of electronic devices (Allsopp, Walters, & Santillo, 2007; Engates & Shipley, 2011).

2.2 Biomonitoring Studies of E-Waste Chemical Exposures From Air, Food, and Water in Local People

Exposure assessment is an essential component of any risk assessment process, and for chemicals released from e-waste recycling sites, biomonitoring studies are required to determine both which chemicals are present and the concentrations found in air, food, and water in human populations living in the area of the recycling sites (Deng, Zheng, Bi, Fu, & Wong, 2007; Fu et al., 2008; Luo, Cai, & Wong, 2007; Wong et al., 2007) to address the issue of chemical dosimetry.

2.3 Epidemiological Studies of Disease Patterns in Recycling Areas in Developing Countries

To assess possible disease patterns at e-waste recycling sites, well-designed epidemiological studies are needed to determine if there are any statistically significant abnormal disease patterns (Chan & Wong, 2013; Leung, Duzgoren-Aydin, Cheung, & Wong, 2008).

2.4 Molecular Epidemiological Studies in Developing Countries Using Omic Biomarkers Measured in Persons Living Around E-Waste Recycling Sites and/or Engaged in E-Waste Recycling

To close the link between measured chemical exposures, human uptakes, and effects, molecular epidemiological studies are needed to "close the linkage" between chemical exposure and biological effect. These molecular epidemiological studies would ideally employ "omic biomarkers" as relative response units for making the connections between documented chemicals exposures and specific biological effects, which could directly be linked to clinical diseases.

2.5 Computational Modeling Studies Integrating Results of Biomonitoring Studies on E-Waste Chemicals With Molecular Epidemiological Studies on the Same Populations

Obviously, these combined types of studies would require sophisticated computational methodologies to integrate analytical exposure data with biological responses as measured by molecular biomarkers. See Fowler (2016) to help identify subpopulations at special risk. As these computational modeling tools become more available in developing countries, it may be possible to transfer the responsibility for conducting such computerized integrative analyses from international public health agencies to risk assessors in public health agencies in developing countries.

References

Afzaal, M., & O'Brien, P. (2006). Recent developments in II–VI and III–VI semiconductors and their applications in solar cells. *Journal of Materials Chemistry, 16*(17), 1597–1602.

Allsopp, M., Walters, A., & Santillo, D. (2007). *Nanotechnologies and nanomaterials in electrical and electronic goods: A review of uses and health concerns.* London: Greenpeace Research Laboratories.

Chan, J. K. Y., & Wong, M. H. (2013). A review of environmental fate, body burdens, and human health risk assessment of PCDD/Fs at two typical electronic waste recycling sites in China. *Science of the Total Environment, 463,* 1111–1123.

Deng, W., Zheng, J., Bi, X., Fu, J., & Wong, M. (2007). Distribution of PBDEs in air particles from an electronic waste recycling site compared with Guangzhou and Hong Kong, South China. *Environment International, 33*(8), 1063–1069.

Engates, K. E., & Shipley, H. J. (2011). Adsorption of Pb, Cd, Cu, Zn, and Ni to titanium dioxide nanoparticles: Effect of particle size, solid concentration, and exhaustion. *Environmental Science and Pollution Research, 18*(3), 386–395.

Fowler, B. A. (2016). *Molecular biological markers for toxicology and risk assessment.* Amsterdam: Elsevier Publishers, 153.

Fu, J., Zhou, Q., Liu, J., Liu, W., Wang, T., Zhang, Q., & Jiang, G. (2008). High levels of heavy metals in rice (*Oryza sativa* L.) from a typical e-waste recycling area in southeast China and its potential risk to human health. *Chemosphere, 71*(7), 1269–1275.

Green, M. A., Emery, K., Hishikawa, Y., Warta, W., & Dunlop, E. D. (2015). Solar cell efficiency tables (Version 45). *Progress in Photovoltaics: Research and Applications, 23*(1), 1–9.

Kahhat, R., Kim, J., Xu, M., Allenby, B., Williams, E., & Zhang, P. (2008). Exploring e-waste management systems in the United States. *Resources, Conservation and Recycling, 52*(7), 955–964.

Kang, H.-Y., & Schoenung, J. M. (2005). Electronic waste recycling: A review of US infrastructure and technology options. *Resources, Conservation and Recycling, 45*(4), 368–400.

Leung, A. O., Duzgoren-Aydin, N. S., Cheung, K., & Wong, M. H. (2008). Heavy metals concentrations of surface dust from e-waste recycling and its human health implications in southeast China. *Environmental Science and Technology, 42*(7), 2674–2680.

Lipomi, D. J., & Bao, Z. (2011). Stretchable, elastic materials and devices for solar energy conversion. *Energy and Environmental Science, 4*(9), 3314–3328.

Luo, Q., Cai, Z. W., & Wong, M. H. (2007). Polybrominated diphenyl ethers in fish and sediment from river polluted by electronic waste. *Science of the Total Environment, 383*(1), 115–127.

Plambeck, E., & Wang, Q. (2009). Effects of e-waste regulation on new product introduction. *Management Science, 55*(3), 333–347.

Streicher-Porte, M., Widmer, R., Jain, A., Bader, H.-P., Scheidegger, R., & Kytzia, S. (2005). Key drivers of the e-waste recycling system: Assessing and modelling e-waste processing in the informal sector in Delhi. *Environmental Impact Assessment Review, 25*(5), 472–491.

Wong, M., Wu, S., Deng, W., Yu, X., Luo, Q., Leung, A., … Wong, A. (2007). Export of toxic chemicals–a review of the case of uncontrolled electronic-waste recycling. *Environmental Pollution, 149*(2), 131–140.

Index

Printed in the United States
By Bookmasters